SEA OF CORTEZ
MARINE INVERTEBRATES

A Guide for the Pacific Coast, Mexico to Ecuador

by

Alex Kerstitch

SEA CHALLENGERS • MONTEREY, CALIFORNIA
1989

A SEA CHALLENGERS PUBLICATION

Copy editor Kenneth Hashagen

First Edition (1989)
Second Printing (1996)

FRONT COVER

Seabutton; *Jenneria pustulata*

Photograph by Alex Kerstitch

Library of Congress Cataloging-in-Publication Data

Kerstitch, Alex N.
Sea of Cortez marine invertebrates.
Bibliography: p.
Includes index.
1. Marine invertebrates – Mexico – California, Gulf of – Identification. 2. Marine invertebrates – Pacific Coast (Central America) – Identification. 3. Marine invertebrates – Pacific Coast (South America) – Identification.
I. Title.
QL138.K47 1989 592.092'61 88-29731
ISBN 0-930118-14-6

SEA CHALLENGERS • 4 Sommerset Rise • Monterey, CA 93940-4112
Printed in Hong Kong through Global Interprint, Peteluma, CA U.S.A.

DEDICATION
To
DONALD A. THOMSON
my
mentor, colleague, and good friend.

FOREWORD

For over a century, amateur and professional naturalists have found the Sea of Cortez one of the most alluring regions in the world. The Gulf exceeds 1,000 km in length, covers nearly 60,000 mi² of ocean surface, and spans nine degrees of latitude, traversing the Tropic of Cancer in its southernmost reaches. The creatures that inhabit its waters arrived there from diverse sources: tropical South America, the Caribbean Sea (before Earth's tectonic forces sealed the Panama Seaway), the cold shores of California (during past glacial periods), and across the vast stretch of the Pacific Ocean from the tropical West Pacific. Biologically, the Sea of Cortez is one of the most productive and diverse seas in the world, harboring an estimated 10,000 invertebrate species (excluding the single-celled protozoans). Given this eclectic backdrop, it is easy to understand the fascination this region has held for naturalists of past and present generations.

The first serious collecting of marine organisms in the Gulf of California was accomplished by John Xantus, a U.S. government tidal observer stationed at La Paz in the 1860's. However, it was not until 1940 that modern marine biology in the Sea of Cortez had its birth with the remarkable pioneering expedition of Ed Ricketts and John Steinbeck aboard the "Western Flyer," a purse seiner out of Monterey, California. The biology and philosophy of this amazing voyage is chronicled in *The Log from the Sea of Cortez* (Steinbeck and Ricketts 1941). Today, much of the mystique in exploring these shores rests on the fact that they are not much changed from the time of Ricketts and Steinbeck. The much heralded Baja trans-peninsular highway has yet to change things on the Peninsula, although it has made travel faster, remote areas more accessible, and a few modern tourist hotels have appeared at Loreto, La Paz, and the Cape region. However, the road is not a high speed motorway, gas is often unobtainable for hundreds of miles, and the long stretches of highway between towns and villages make travel an adventure of the highest sort. The drive from Tijuana to Cabo San Lucas traverses more than 1,000 miles of largely uninhabited coastal, desert, and mountain backcountry. Hundreds of dirt roads peel off the two-lane highway, allowing modern adventurers to penetrate into regions as remote as they care to explore.

The 1950's and 60's ushered in an era of organized research expeditions to the Sea of Cortez and Baja California. These expeditions resulted in voluminous publications describing the flora and fauna of the Gulf and Baja Peninsula. Despite this invaluable baseline taxonomic work, an estimated 60% of the invertebrate species of the Gulf and Baja Peninsula remain undescribed. The region is still fertile ground for naturalists inclined to investigate unknown faunas.

By the late 1960's, the focus in biology began to shift from descriptive natural history to quantitative and experimental biology — what today's university student calls "modern ecology." This shift in research has been slow to materialize in the Sea of Cortez, largely due to its remoteness and our limited basic knowledge of its flora and fauna, and the major research emphasis continues to be on collection and documentation of the fauna. One of the first modern ecological studies in the Gulf was Robert Parker's analysis of subtidal benthic invertebrate communities (Parker 1962). Parker identified 1,150 species of invertebrates from 200 trawl collections taken in the Gulf. In 1966, Robert Paine published his now classic hypothesis of predator regulation of community species diversity, based in part on his study of Gulf intertidal communities. Parker and Paine, like Steinbeck and Ricketts, were struck by the extreme richness of the Gulf's invertebrate fauna. Parker claimed that no other benthic communities in the world could compare in species richness to those of the Sea of Cortez. The same observation was expressed by noted coral reef biologist, Sir Maurice Yonge, after a few hours of tidepooling on a limestone reef in the northern Gulf in 1970. Paine, Parker, and other biologists have also been impressed by the tremendous number of vertebrate and invertebrate predators that occur at all depths in the Gulf.

Following close on the heels of Parker and Paine's studies, the first ongoing marine science program in the Gulf was established by a U.S. institution, at the University of Arizona. This program continues today and has trained and graduated scores of students whose research has shed light on the natural history and ecology of the Gulf. Much of this work is summarized in two benchmark texts, one on the invertebrates (Brusca 1980) and one on the fishes (Thomson, Findley, and Kerstitch 1979) of the Gulf. Alex Kerstitch has been associated with the University of Arizona Marine Science Program since its inception and has photographed and studied marine life in the Sea of Cortez since the early 1960's.

A great deal remains to be done before the nature of animal and plant communities in the Sea of Cortez is understood, although at this point some rough generalizations can be made. Except for a few common forms, the invertebrate *species composition* (the species that comprise a local community at any given time) of most Gulf intertidal and shallow-water areas is largely unpredictable and under the influence of a complex network of interacting physical and biological elements. However, the *relative species richness* (the number of species present in any given area) is reasonably predictable and largely a function of local substrate types and variety. In the Gulf, species richness is typically highest on relatively stable beaches and shallow bottoms composed of soft eroded rocks such as limestone and sandstone and lowest on beaches composed of smooth hard rocks such as basalt and highly unstable beaches of coarse sand or cobble. Bottoms that have a variety of substrate types harbor more kinds of invertebrates than do more homogeneous bottoms. Simply put, the reason for these relationships is that stable but heterogeneous substrata provide a greater variety of "niches" or microhabitats among which different kinds of invertebrates can coexist.

On a larger scale, high species diversity in the entire Gulf region is largely due to two phenomena: (1) the great variety of general habitats that occur in the Gulf including mangrove lagoons, coral communities and reefs, shallow and deep basins, and a great variety of shore and subtidal substrate types; and, (2) the complex geological/oceanographic history of the Gulf, including its past invasions of animal immigrants from north, south, east, and west.

Invertebrate communities in the Sea of Cortez are also influenced by seasonal oceanographic conditions, especially in the northern Gulf, where marked seasonal changes in water temperature and wind pattern occur. The Gulf is unusual because it is a single enclosed body of water that possesses two rather different ecological regions. The shallow northern Gulf, north of the midriff islands (Islas Angel de la Guarda and Tiburón), is strongly influenced by the climate of the surrounding Sonoran Desert and experiences extreme annual variations in sea water temperature. As a result, the northern Gulf is essentially a warm-temperate marine environment during the winter, but a subtropical marine environment during the summer. In this region, distinct seasonal changes occur with respect to the invertebrates and algae as certain tropical species disappear during the cold winters and other, temperate species are lost during the warm summers. The southern Gulf, with its greater area, numerous deep basins, and proximity to the open Pacific, is more under the influence of oceanic conditions and is largely a stable subtropical environment year round. Research to date suggests that the northern Gulf intertidal and shallow-water environment tends to be more *physically controlled,* while the southern Gulf intertidal and shallow-water environment tends to be *biologically accommodated.* In the former case, community composition is largely regulated by physical aspects of the variable environment, which periodically cause marked physiological stress on the species that are present. In the latter case, more stable environmental conditions result in communities that are "in balance" and largely regulated by biological interactions among species.

As ecological knowledge about the Sea of Cortez is gained, more attention must be given to problems of marine pollution and the loss of vitally important coastal habitats. Destruction of the Gulf's lagoons and mangrove esteros by pollution and commercial development (e.g. witness the recent devastation of Estero Algodones by Club Med, near Guaymas) is slowly eliminating these valuable habitats, which offer shelter, food, and breeding sites for a multitude of migratory birds and marine life, and also play a critical role as nursery grounds for commercial fisheries species (fish, shrimp, and molluscs). It is likely that increasing pressure from Mexico's bureau of tourism development (FONATUR) will ultimately open the way to development and unavoidable damage to unique habitats such as the Pulmo Bay coral reef, south of La Paz. FONATUR has already targeted the isolated, pristine shores of Puerto Escondido for a joint Mexican-

French resort development patterned after Cancun and Cannes. And if intense internal pressure to develop the coast weren't enough, international pressure is also growing. Japanese and Korean fishing boats, along with the Mexican fishing fleet, now drag nets across the Gulf sea floor. On shrimp boats, 90% of the catch is fish and invertebrates that are unused and die in the net or on deck before being discarded overboard. Israeli engineers are helping to develop Mulege, West Germans are experimenting with solar-powered desalination in La Paz, and American companies are waiting in line for permits to open hotels and manufacturing plants along the shores of the Sea of Cortez.

The key to understanding, appreciating, and conserving any natural region is knowledge of what species are present and the role they play in the ecological drama of the area. Alex Kerstitch's field guide provides access to much of this information. With it, any amateur or professional can easily identify many of the common intertidal and subtidal invertebrates of the Gulf of California. Alex Kerstitch's skills as a diver, photographer, artist, and naturalist are a unique combination of qualities that have blended to create a wonderful text — a combination of art and information, a book both useful and pleasing to the eye.

Richard C. Brusca
Chairman, Department of Marine Invertebrates
San Diego Natural History Museum

ACKNOWLEDGMENTS

A book of this nature cannot be realistically completed by one author without the collective assistance of a number of individuals. The contributions of each, whether major or minor, is sincerely appreciated.

First and foremost, I am deeply indebted to three contributors who authored major sections: Dr. Hans Bertsch (opisthobranchs and cnidarians), Mr. Ron H. McPeak (cnidarians), and Dr. L. Yvonne Maluf (echinoderms). I also thank Dr. Richard C. Brusca for his suggestions while reviewing the manuscript and for preparing the isopod accounts, and Janet Voight for the octopod accounts.

A number of specialists have contributed specimen identification and systematic review of controversial and difficult groups. I greatly thank Dr. Aisla M. Clark (asteroids), Dr. Maureen E. Downey (asteroids), Dr. L. Michael Dungan (barnacles and sponges), Dr. Wesley M. Farmer (opisthobranchs), Dr. John S. Garth (brachyurans), Dr. Terrence M. Gosliner (opisthobranchs), Dr. Janet Haig (anomuran crustaceans), Dr. Gordon Hendler (echinoids and holuthuroids), Dr. Eric Hochberg (octopods and squids), Dr. Jerry Kudenov (worms), Dr. Alberto Larrain (echinoderm taxonomy and ecology), Dr. Raymond B. Manning (stomatopods), Dr. James H. McLean (gastropods), Dr. James Nybakken (cone snails), Dr. David L. Pawson (echinoids and holothuroids), Dr. Peter E. Pickens (anemones), Roy Poorman (molluscs), Carol Skoglund (molluscs), Dr. Mary K. Wicksten (decapod crustaceans), and Dr. Austin Williams (scyllarids).

This book would not have been possible without the help of my mentor, associate, and good friend, Dr. Donald A. Thomson, who patiently guided me and provided emotional support during my formative and academic years at the University of Arizona and in the Gulf of California.

A number of people assisted me in various ways; their contributions are warmly appreciated. I thank Peggy and Rick Boyer, Felipe Maldonado, Jack Abert, Chuck and Gratia Duecy, Dr. Matt Gilligan, Dr. Lloyd T. Findley, Dr. Steve Schuster, Dr. Alan Harvey, Dr. Phil Hastings, Dr. Chris Peterson, Marc Parr, Dr. Manuel Molles, John Van Ruth, Carol Enevoldsen-Madachy, Matt Ankley, David Berwick, Sandy Richie, Lauro Velarde, Sra. Mary and Sr. Enrique Ricaud Sr. and Jr., Dr. Michael T. Ghiselin, Elissa Lenard, Tom Smith, Luis E. Aguilar Rosas, Guillermo Ballesteros, Joy Wissinger.

The joy of writing this guide would have been incomplete without the participation of my field and diving companion, moral supporter, and endearing wife, Dr. Myra M. Kerstitch.

TABLE OF CONTENTS

INTRODUCTION

In recent years a number of natural history books on the Gulf of California (Sea of Cortez) have appeared as the result of increasing interest by visitors discovering the richness of its waters. Fishermen, divers, beachcombers, and students of the sea are eager to identify and learn more about the creatures they encounter along Gulf shores. Two books, *REEF FISHES OF THE SEA OF CORTEZ* (Thomson, Findley, and Kerstitch 1979) and *COMMON INTERTIDAL INVERTEBRATES OF THE GULF OF CALIFORNIA* (Brusca 1980) are the most comprehensive on the subjects they cover. Not only are these texts helpful for their broad coverage, but they also stand out because of the scholarly treatment given to each species.

The main objective of the present guide is to provide a source of color photographic identification for both the common and uncommon invertebrates of the Gulf of California. It should both supplement and complement Brusca's book.

Based on the literature and my personal observations of the invertebrate fauna of the Gulf, I provide information on habitat, depth, range, and distribution of each species treated here. These data are, of course, subject to future revision as new information becomes available. The depth ranges of many invertebrates vary seasonally. Some animals, such as the Spanish shawl, *Flabellina iodinea,* and the Lucas cleaner shrimp, *Periclimenes lucasi,* occur in shallow water during certain periods and deeper water at other times. Or, they may be entirely absent during certain seasons, particularly in areas where sea temperatures vary considerably. For example, onshore sea surface temperatures in the upper Gulf may fluctuate from 50° F (10° C) in winter to over 90° F (32° C) in summer, strongly affecting the seasonal abundance of many shallow water invertebrates. Although a number of species treated here will be encountered only subtidally by divers, many will be found intertidally as well; others will be associated exclusively with the intertidal zone.

The majority of marine animals from the Gulf do not have accepted common names and are usually referred to only by their Latin scientific names. To accommodate users of this guide I have given each species a vernacular name. In most cases their names refer to some notable attribute of the animal such as size, color, abundance, rarity, or some unique behavioral trait. Thus, the peacock mantis shrimp, *Hemisquilla ensigera californiensis,* is so named because of the large colorful eyespots on its tail, and the keeled sea star, *Asteropsis carinifera,* because of a distinct ridge of spines running down the center of each arm.

In order to provide an accurate morphological description of each species the use of technical terms cannot be avoided. I have tried to keep these to a minimum and include definitions in the *GLOSSARY* or directly in the text. In most cases I do not give overall color description of a species when the color photo provides an accurate representation of the live animal. Distinctive color patterns are described only when necessary to distinguish between closely related species.

The upper Gulf is characterized by extreme tidal ranges, and low tides can expose large stretches of rocky reefs where intertidal animals are easily observed in tide pools. In the spirit of conservation, intertidal and subtidal animals should be left undisturbed and only dead specimens, such as empty sea shells, should be collected. Above all, overturned rocks should be returned to their original positions. Anyone wishing to capture edible animals, such as crabs, clams, or scallops, should first contact local fishing authorities to inquire about fishing regulations. Fishing licenses are usually required to capture certain marine animals, particularly the commercially harvested species such as lobsters, rock scallops, and the Cortez conch.

I hope this guide will not only allow the user to identify an animal at a glance from the color photographs, but will also whet the appetite to learn more about the Gulf's rich and unique marine fauna.

HOW TO USE THIS BOOK

Organization of the text

The organization of the invertebrate phyla is a phylogenetic sequence, starting with the simplest organisms and ending with the most advanced. The order of species within each of the eight phyla represented in this book is generally based on *Common Intertidal Invertebrates of the Gulf of California* (Brusca 1980) and *Intertidal Invertebrates of California* (Morris, Abbott, and Haderlie 1981).

Each species account includes the following information:

Species name:

The scientific names given here are based on the most recent authoritative literature and, in most cases, have been provided by biologists specializing in particular groups of Gulf invertebrates. Whenever the identification of particular species proved to be difficult or altogether impossible during the preparation of this book, the animal was collected, and both specimen and color photo were sent to appropriate specialists for identification.

In several instances positive identification was not possible for a number of reasons: 1) The preserved specimens were not available for study and identification was risky from color photos alone. 2) The systematics of a particular species or group, such as gorgonians, has not been well studied. 3) The animals represented are proposed new species and are currently being described.

A number of generic and specific names have been changed in recent years, and in these cases both the former and current names have been provided.

Common names:

The English common name is, whenever possible, descriptive of the animal. A few accepted common names of Gulf animals have appeared in the literature and are herewith used, e.g.: Sally Lightfoot crab, *Grapsus grapsus;* California spiny lobster, *Panulirus interruptus;* Mexican dancer, *Tridachiella diomedea;* tent olive, *Oliva porphyria;* sulfur sponge, *Aplysina fistularis;* crown-of-thorns sea star, *Acanthaster;* man-of-war, *Physalia;* but since the majority of the animals treated here have no standardized vernacular names, new ones have been provided.

Description:

Although overall body shape and color may be enough to identify some invertebrates, a listing of diagnostic characters is also provided to positively identify each animal and to separate them from closely related species. Technical terms are defined in the glossary, but whenever possible, familiar terms are used in the text.

Size:

The size range given represents the smallest to largest specimens that may be encountered, the average size lying somewhere in between. Metric equivalents follow measurements in inches and feet; the metric conversions are approximate.

In a number of animals standard measurements are specifically designated because of their irregular body anatomy. Thus, the size of a sea star is given by its arm radius, a crab by its carapace length, and a hard coral head by its diameter. Others such as worms, sea slugs, or sea cucumbers are simply given as total body length.

Habitat:

Most invertebrates are associated with specific environmental conditions, or habitats. Some live in rocky substrates, under rocks, or in crevices, while others may occur exclusively in sandy or muddy bottoms. Many organisms, however, prefer to live in close association with other animals. For example, ovulid cowries live exclusively on gorgonians, some isopods are parasitic on fishes, and a number of shrimps are associated strictly with cnidarians or echinoderms. The habitat information provided here is based on recent publications and my own observations.

Although some organisms occur only intertidally, many occur throughout a wide range of depths.

Distribution:

The Gulf invertebrate fauna is characterized by a mixture of species of wide-ranging origins; some have tropical affinities (Panamic, Indo-Pacific, West Indies), others are temperate (Californian), and a number of species are endemic to the Gulf, that is, they do not occur outside the Gulf.

The Gulf has been divided into three faunal subdivisions, each characterized by distinctive faunal communities: the northern, central, and southern Gulf.

The northern Gulf is a region extending from the mouth of the Colorado river to just north of Kino and across the Gulf to Bahía San Francisquito on the Baja California coast.

The central Gulf is the region beginning at Bahía Kino (including the midriff islands) south to Guaymas, and across the Gulf from Bahia San Francisquito south to La Paz.

On the mainland side, the southern Gulf begins just below Guaymas, Sonora, south to Mazatlán, Sinaloa, and across the Gulf, extending from La Paz south to Cabo San Lucas at the tip of the peninsula.

The geographical range of most invertebrates treated here extends beyond the Gulf. The majority range south into the eastern Pacific tropics (south to Peru). The distributional limits of few species extend outside the Gulf as far north as Washington, while others may range southward all the way to Chile. The geographical range of a few species extends to the Indo-West Pacific, or the western Atlantic.

The distribution of a number of species outside the Gulf is poorly known, or has not been recorded.

Remarks:

Although little is known about the natural history of most Gulf invertebrates, this section provides available information pertaining to general behavior, feeding habits, abundance, reproduction and life cycle, ecological roles, and conservation status. When appropriate, remarks also indicate whether a species is venomous and should be handled with care.

Porifera

Because of their amorphous appearance, sponges are occasionally confused with plants. They are, however, simple multicellular animals that vary in shape from large vase-like forms to small, flat encrustations covering rocks. The porous body pumps water through a filtering system made up of a network of channels lined with specialized cells. As water passes along the channels, minute food particles are collected, oxygen is removed, and waste products are excreted. The skeletons of sponges are made up of either tough spongin fibers, spicules (calcareous or siliceous), or a combination of both.

The great variety of form and size among sponges is often related to their immediate habitat. Where plankton-rich currents are swift, sponges reach maximum development. Some species can grow to be massive and tubular under optimum conditions, but only develop into simple encrustations in less favorable conditions. This is one of the reasons why sponges are difficult to identify and are poorly known in the Gulf of California. A key to 22 species of common intertidal sponges of the Gulf is provided by Brusca (1980).

Cnidaria (Coelenterata)

Cnidarians are members of a rather large phylum consisting of two morphological types, the medusa (jellyfishes) and the polyp (corals, sea anemones, gorgonians, and allies). Both body forms are radially symmetrical, with a single body opening that serves as mouth and anus. This opening is surrounded with tentacles armed with stinging cells. These venomous cells, or nematocysts, are used primarily to paralyze and capture prey. The presence of nematocysts is one of the most distinctive characteristics of cnidarians.

The three classes recognized include: (1) Hydrozoa (hydroids, hydromedusae, siphonophorans, etc.), (2) Scyphozoa (true jellyfishes), and (3) Anthozoa (anemones, corals, and their allies). In the hydrozoans there are usually both medusae and polyp body forms. Some species of medusae are among the most venomous of marine animals. The Pacific man-of-war, *Physalia utrilculus,* is perhaps the best know of all Gulf of California medusae that drifts inshore in large numbers during certain times of the year.

In the scyphozoan class, the medusa is the dominant body form, consisting of a firm, jelly-like layer in the body wall and nematocyst-laden tentacles. They feed mostly on small planktonic organisms by capturing them with their stinging tentacles or by trapping them in mucus contained in the upper portion of the bell. The notorious and highly venomous sea wasp, *Chironex fleckeri,* of Australia is a member of this class, although some authors prefer to place this dangerous animal in a separate class, Cubozoa. There are a number of cnidarian species in the Gulf of California that are mildly toxic to swimmers and divers.

All anthozoans have polyp body types and are usually anchored to hard surfaces. Many species tend to form colonies, particularly anemones and corals. Some anthozoans form partnerships with unicellular algae known as zooxanthellae that reside in the polyp tissues. The zooxanthellae provide oxygen and photosynthetic products to the host and, in turn, receive nutrients and carbon dioxide.

Platyhelminthes

Known as flatworms, members of this phylum have a simple, bilaterally symmetrical body plan. They are compressed and many are delicate and prone to tearing if not handled carefully. Two of the three classes, tapeworms and flukes, are entirely parasitic, infesting fishes and occasionally crustaceans. Only the polyclad flatworms (order Polycladida) are treated here. These often colorful marine worms are relatively large and are characterized by a flat, cilia-covered body that has an incomplete digestive tract.

Polyclads are usually found on rocky substrates, crawling on or hidden under the rocks. When disturbed by predators, some species have the ability to swim by undulating the margins of the body. They are mostly carnivorous, feeding on small polychaete worms and other small invertebrates. Tropical species are usually brightly colored with striking patterns.

Nemertea

Nemerteans are popularly called ribbon worms because of their slender and often extremely long bodies that range in size from a few millimeters to 10 meters in one particular genus, *Lineus,* from the eastern Atlantic.

Although related to flatworms, ribbon worms are characterized by having a complete digestive system with both a mouth and an anus. One of the most important distinguishing characteristics of ribbon worms is the food-gathering proboscis that can be everted to capture prey. Armed with spines and/or an adhesive mucus, the proboscis is quickly thrust out to trap the prey. They feed on other worms, crustaceans, molluscs, and sometimes small fishes. The mucus in some ribbon worms is venomous and is used to paralyze prey. In others, the proboscis is armed with stylets, or minute arrows used to harpoon their victims. Approximately 14 species have been reported intertidally from the Gulf of California. The zebra worm, *Baseodiscus mexicanus,* is one of the most common intertidal and subtidal Gulf species.

Annelida

This diverse group includes three classes — earthworms, leeches, and marine polychaetes — all of whom are characterized by their segmented bodies. These segmented worms can dig and burrow into various substrates more efficiently than non-segmented worms. Of the three annelid classes, only the polychaetes are treated here. This class can be roughly divided into two ecological groups: a) free-living or errant and b) sedentary, those worms that burrow or are tube dwelling.

Most polychaetes have well-developed digestive, nervous, and circulatory systems. They are equipped with paddle-like appendages called parapodia, which function in gill support and locomotion. In some species the parapodia possess numerous long bristles. The bristles are sometimes detachable and are associated with a venom that can be mildly to highly toxic.

Many polychaetes are tube dwellers. Some construct parchment tubes while others form calcareous tubes that may have calcareous plugs. Many tube-dwelling species, such as feather duster worms, possess colorful tentacles that extend from the entrance of the tubes. Feather dusters use the tentacles for respiration and for filter feeding. They are common throughout the Gulf and are usually found on rocks, docks, boat hulls, or any hard, submerged surface.

Sipuncula

Sipunculans are unsegmented, cylindrical, worm-like animals that are exclusively marine. They possess a mouth, surrounded by small tentacles, located terminally on a protrusible proboscis. Their bilaterally symmetrical bodies are often covered with short spines or papillae and they are frequently ribbed due to the underlying muscle layers. When contracted they resemble a small peanut, hence the common name "peanut worms."

Many sipunculans feed on organic material filtered from large quantities of ingested sand or mud. Those species equipped with particularly long ciliated tentacles may filter organisms directly from the water.

Most peanut worms are found intertidally and subtidally, buried in sand, mud, or under rocks. Others may inhabit empty shells, rock crevices, porous reefs, or sponges.

Mollusca

Next to arthropods this phylum is the second largest both in number of species and diversity of form. There are approximately 100,000 living species in seven classes, of which four major classes are treated here. They include chitons, snails, bivalves and cephalopods. At first glance these groups bear little resemblance externally to one another. The octopus and scallop, for example, superficially appear unrelated, yet they have a fundamentally similar body plan. Because of evolutionary pressures certain morphological characteristics have been modified to meet the specialized living requirements of these animals.

As varied as they may appear, all molluscs are bilaterally symmetrical animals with a head, foot, and a visceral mass containing the internal organs covered by a thick sheet of skin called the mantle. One feature unique to this phylum is the radula. Most groups are equipped with one or more of the chitinous, toothed ribbon that are used for scraping food material from the substrate. In some molluscs, however, radular teeth have been modified for drilling, biting, or stinging in order to attack and capture their prey. A few groups include species that have a venom apparatus associated with the radula that is used to paralyze prey. Cone shells, augers, turrids, and octopods fall into this category. Some species are even dangerous to humans and several mortalities have resulted from the careless handling of some molluscs, such as the geography cone of the Indo-West Pacific and the blue-ring octopus of Australia.

Molluscs are well represented in the Gulf, where they have exploited nearly every conceivable habitat from the intertidal zone down to depths of several thousand feet.

Arthropoda

This phylum is the largest and most diverse of all invertebrate groups. Over 1,000,000 living species are known, although most of them are insects. The class Crustacea is one of the most speciose aquatic groups, comprising more than 35,000 species represented. A few taxonomists have suggested this class be placed in its own phylum, but I have followed the traditional classification and retained it as a class of the arthropod phylum.

The most distinctive feature of the crustaceans is the hard, segmented exoskeleton that protects the internal soft parts. This articulate armor is usually divided into a head, thorax, abdomen, and tail, but in some groups the head and thorax are fused together to form a single structure known as the carapace. The hard, chitinous exoskeleton cannot grow, so it must be shed periodically to allow internal body growth. During molting, a thin section between plates on the thorax, or between the thorax and abdomen splits and allows the animal to escape in its newly formed exoskeleton.

The head is equipped with two pairs of sensory antennae, eyes, and three to five pairs of mouth parts. Legs are attached to the thorax, and often bear claws used for crushing or cutting food material.

Ten distinct groups are treated here. They are: 1. barnacles (Cirripedia), 2. isopods (Isopoda), 3. mantis shrimps (Stomatopoda), 4. true shrimps (Natantia), 5. lobsters (Palinura), 6. ghost shrimps (Thalassinoidae), 7. porcelain crabs (Porcellanidae), 8. hermit crabs (Coenobitoidea and Paguroidea), 9. mole crabs (Hippoidae), 10. regular crabs (Brachyura).

Echinodermata

Most authorities believe echinoderms to be a transitional group between invertebrates and chordates. At first glance, however, they do not seem to be very advanced. They have no head and their nervous, digestive, excretory, and respiratory systems are simple. The phylum includes sea stars, sea urchins, brittle stars, and sea cucumbers and their allies. Unlike arthropods, they possess an internal skeleton made of calcareous plates which are covered with skin. The surface is often covered with spines.

The secondary radial symmetry that characterizes echinoderms is one of the distinctive external characters of this phylum. More specifically, this symmetry is pentamerous in most echinoderm groups, meaning that their body plan consists of five more or less equal parts. This is particularly evident in sea stars and brittle stars.

An important and unique internal echinoderm character is the hydraulic water vascular system responsible for tube-feet movement. Tube-feet function in locomotion, feeding, respiration, and sensory activities.

Echinoderms are found in nearly all habitats and occur from the shallow intertidal zone to the deepest parts of all seas. They employ a great variety of feeding mechanisms, including suspension and detrital feeding, as well as herbivory and carnivory.

PICTORIAL KEY TO PHYLA

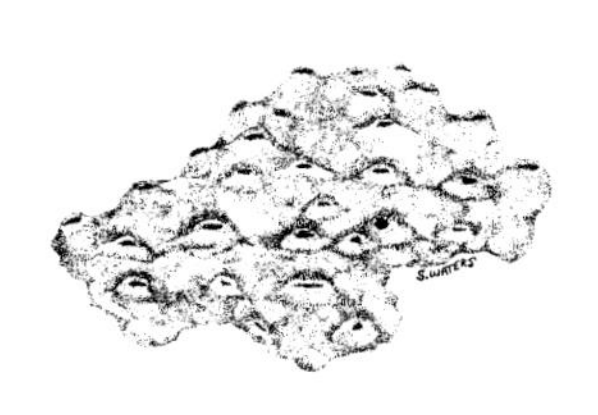

1. Sponges P. 15

2. Hydroids P. 18

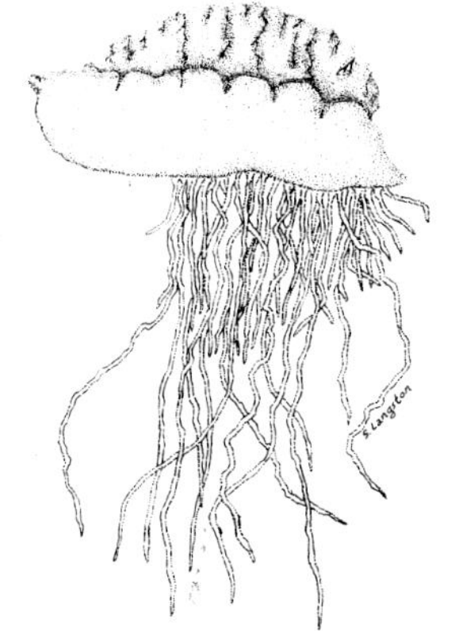

3. Jellyfishes P. 20

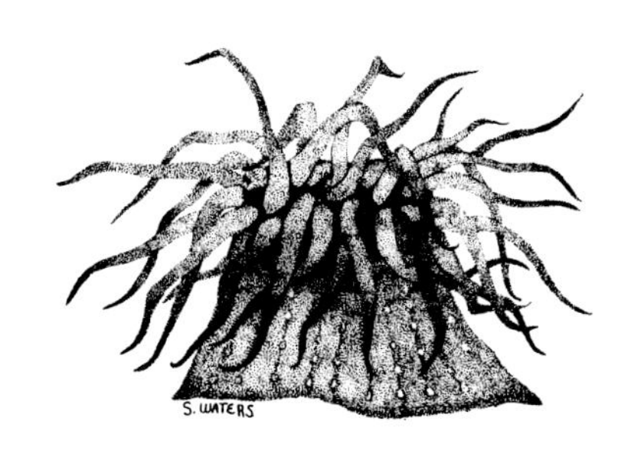

4. Anemones P. 21

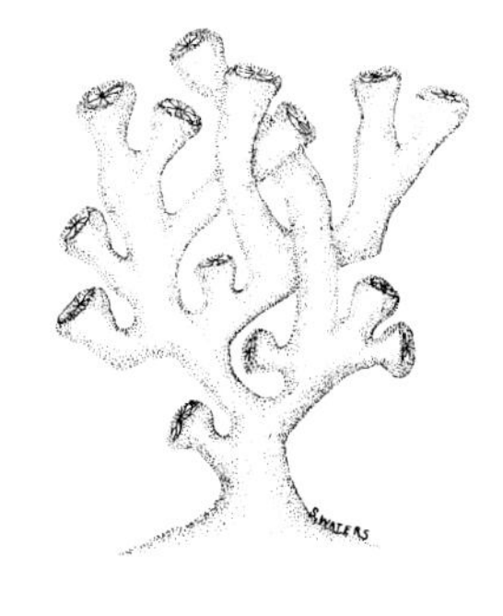

5. Corals P. 26

6. Gorgonians P. 29

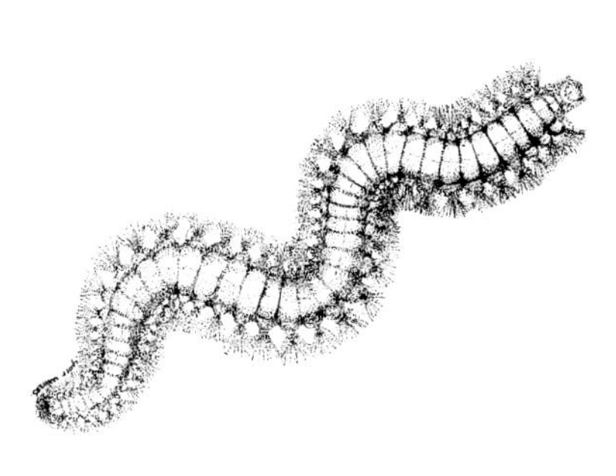

7. Polychaete Worms P. 32

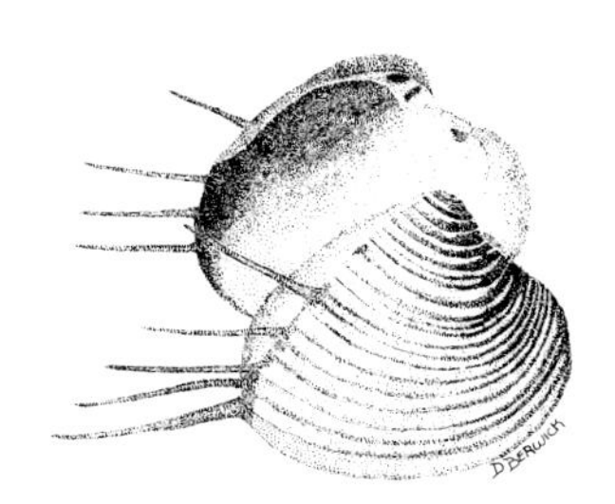

8. Clams P. 34

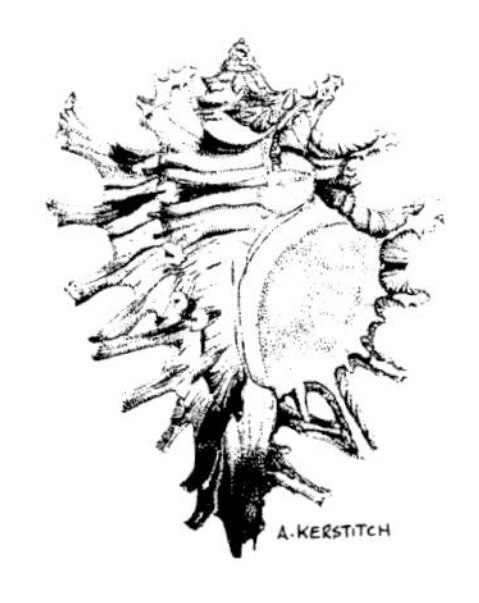

9. Snails P. 37

10. Nudibranchs P. 54

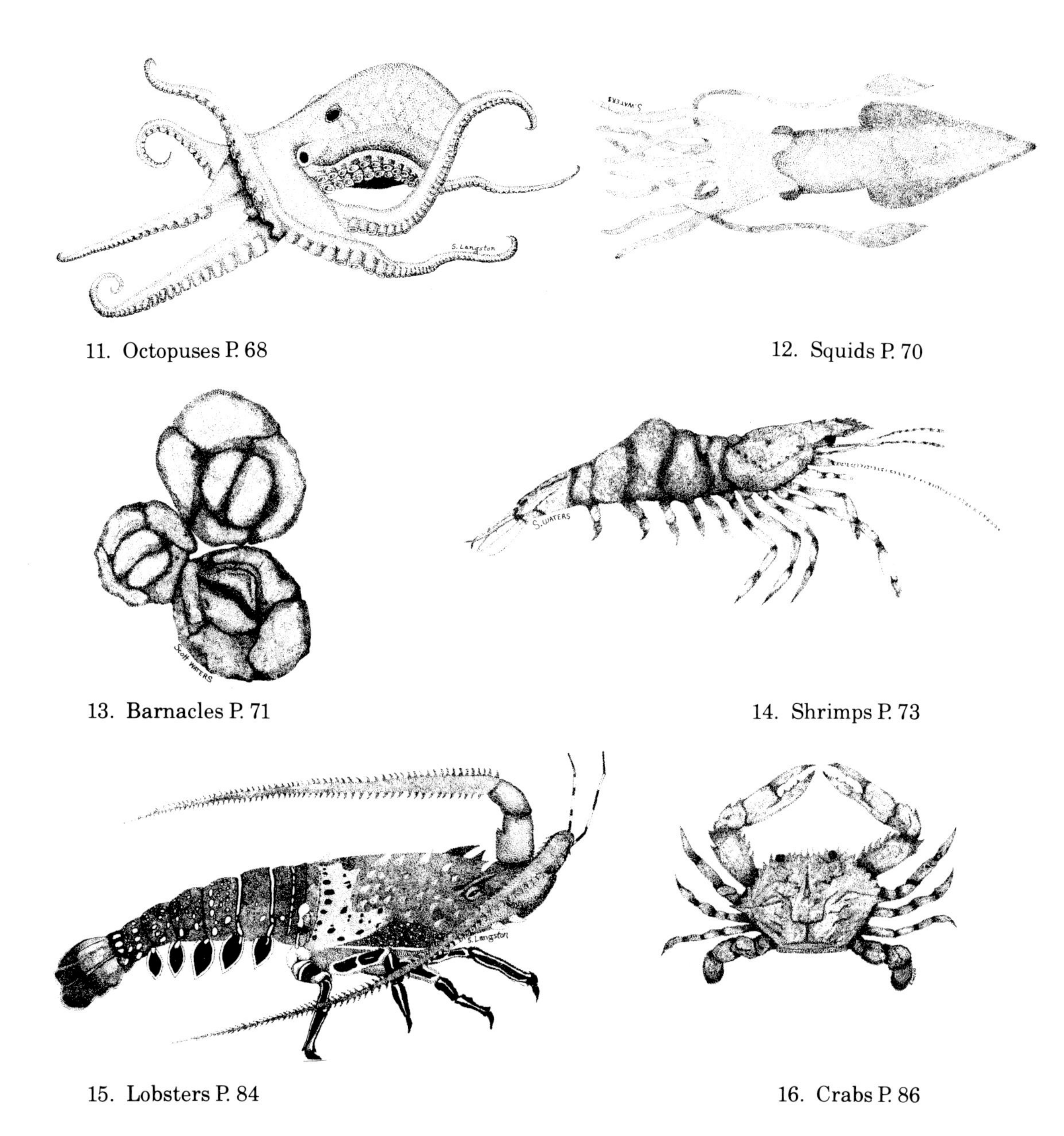

11. Octopuses P. 68

12. Squids P. 70

13. Barnacles P. 71

14. Shrimps P. 73

15. Lobsters P. 84

16. Crabs P. 86

(Drawings 1, 4, 5, 7, 12, 13, 14, 16, 19 by Scott Waters; 3, 11, 15, 18, 20, 21 by Scott Langston; 2, 8, 10 by David Berwick; 6 by Sean Coldiron; 9 by Alex Kerstitch; 17 by Chris Wendling.)

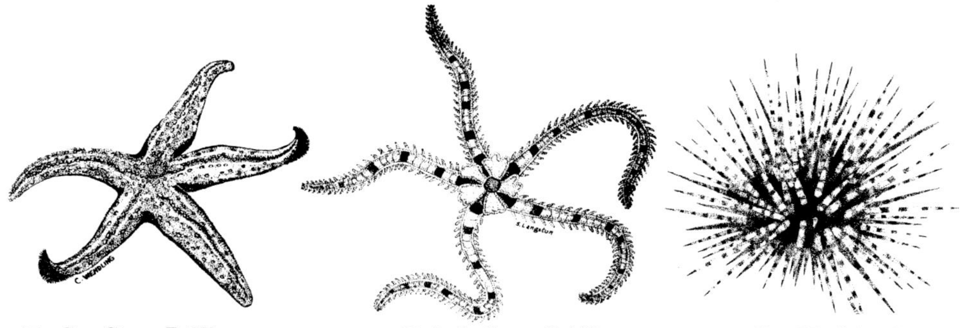

17. Sea Stars P. 97

18. Brittle Stars P. 102

19. Sea Urchins P. 104

20. Sand Dollars P. 106

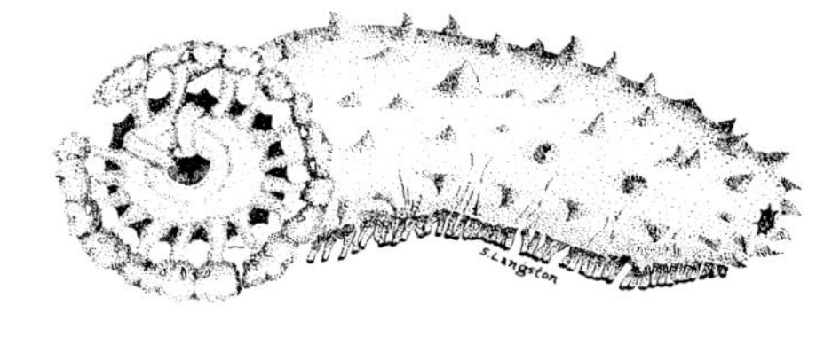

21. Sea Cucumbers P. 108

GLOSSARY

aboral — A region of the body opposite the mouth.

acontia — Threadlike structures that may be shot out of the column of a sea anemone and often bear nematocysts.

actinopharynx — Entrance into coelenteron (gut cavity) of anemones.

agassizin — A biological chemical produced by *Hypselodoris agassizii,* probably used for defense.

ambulacra — Radiating grooves on the legs or body of echinoderms where podia (tube feet) of water-vascular system project to the outside.

amphi-American — A pattern of geographic distribution, wherein a species occurs on both sides of America (in the Pacific and the Atlantic).

amphinomid — Polychaete worms of the family Amphinomidae.

amphipods — Minute crustaceans of the order Amphipoda: includes sand hoppers, sand fleas, etc.

antennules — The first pair of antennae on the head of a crustacean.

anterior — The head or front end of an organism.

anterior canal — In snails, the front end of the aperture (opening in the last whorl).

aperture — The main opening into the first (largest) whorl of a gastropod shell.

apex — The point at the tip of the spire of a gastropod.

apical orifice — An opening at the top or apex of some organisms, such as the keyhole limpet.

appendage — Any member of the body diverging from the axial trunk, such as limbs or antennae.

arborescent — Tree-like in appearance.

arminid — Nudibranchs of the suborder Arminacea.

asteroids — Sea stars.

autotomy — Self-amputation.

axial ridges — Ribs or ridges which are longitudinal, or parallel to the axis of coiling on a gastropod.

bifurcate — To divide or split into two.

biosynthetic — Chemical compounds produced by living organisms by either synthesis or degradation.

body whorl — The basal and usually the largest whorl of a gastropod, or univalve shell (snail).

branchial crown — Respiratory tentacles, particular of feather duster or fan worms.

branchial lobes — Roundish respiratory projections on various invertebrates.

bryozoans — Small aquatic animals of the phylum Bryozoa; grow as encrusta-

bivalve — Any mollusc having two valves or shells hinged together.

tions, or tuft-like or moss-like aggregate masses.

buccal — Pertaining to the mouth or cheek.

byssal threads — Attachment threads of bivalve molluscs.

calcareous — Composed of calcium carbonate or chalk.

capitate tentacles — Short tentacles with a terminal knob, usually studded with nematocysts, or certain anemones.

carapace — A chitinous shield covering the head part of the body of certain crustaceans.

caprellids — Amphipod crustaceans of the family Caprellidae.

carina — A prominent keel or keel-like ridge on the body or on an appendage.

cephalic tentacle — A fleshy and flexible appendage on the head of an organism.

cerata — Processes on the backs of some shell-less marine molluscs (sea slugs) that are used for respiration and other purposes.

chela — The pincerlike claw on the tip of some limbs of certain crustaceans.

cheliped — One of the pairs of legs (appendages) that bears the large claw in decapod crustaceans.

chemosensory — Any sense organ responding to chemical stimuli.

chitinous — A hard, amorphous substance that makes up the covering of crustaceans, the internal shell remnants of squids, and the horny operculum of many gastropods.

chlorophyll — Green pigment found in plants and some animals; necessary for photosynthesis.

chromatophore — A pigment cell, usually in the epidermis or dermis.

cilia — Microscopic hair-like processes that are often used in locomotion and food-gathering.

circumtropical — Surrounding or distributed throughout tropical regions of the world.

cirrus (pl. = cirri) — Small flexible tentacle-like sensory appendage.

cnidarians — Coelenterates.

coelenterates — Any animal of the phylum Cnidaria (also known as Coelenterata) including anemones, corals, sea fans, jellyfishes, and others.

columella — The central pillar of a univalve shell (gastropod) around which the whorls are built.

column — A stalk connecting the base and disc of anthozoans (e.g. sea anemones).

commensal — The relationship between two different species in which one obtains food, protection, or other benefits without damaging the other.

convoluted — (convolute) — To twist, or roll up.

copepods — Minute crustacean of the subclass Copepoda.

corallites — The skeleton of an individual coral polyp.

corbulae — Reproductive structures of certain hydroids, such as *Aglaophenia*.

cortex — The outer layer of a structure.

cosmopolitan — Widely distributed over the globe.

crown — Top or highest point.

crustacean — Any chiefly aquatic arthropod of the class Crustacea.

cryptic — Serving to conceal the pattern or coloring of an animal.

cusps — A point, especially on the crown of a tooth or denticle.

Cuvierian tubules — Sticky tubules expelled from the anus of sea cucumbers as a defensive reaction.

dactyl — Terminal or distal article of a crustacean leg or other appendage.

dactylozooid — A tentacular mouthless zooid in certain hydrozoans that performs tactile and protective functions of the colony.

dendritic — Branching in form, tree-like.

depauperate — Imperfectly developed or impoverished.

desiccation — Process of drying up.

detritivore — Organism that feeds on animal waste and dead organic matter.

dichotomous — Regularly dividing by pairs; successive bifurcation; two forked.

dimorphism — Existence within a species of two distinct forms according to color, sex, size, organ structure, etc.

distal — Relatively remote from the center of the body or point of attachment.

diterpene — Organic hydrocarbon compound.

dorsomedial — Located towards the back and near the midline.

ecophenotype — A phenotype (body form) modified by specific adaptive response to environmental conditions.

ectoderm — The outer cellular membrane of an early embryo.

El Niño — The cyclic phenomenon that involves the widespread warming of the equatorial Pacific ocean and results in the disruption of marine ecosystems and weather patterns throughout the world.

endemic — Restricted to a particular region.

epipelagic — Oceanic zones, in near-surface waters where enough light penetrates for photosynthesis.

epizoanthid — Small solitary anemones of the genus *Epizoanthus*.

exoskeleton — External skeleton or supporting covering of animals (particularly arthropods).

extrabranchial — Located outside the branchial arches (particularly in fishes).

fasciole — A band of oar-shaped, ciliated spines used to produce currents along the body surface of heart urchins.

filiform — Filament or thread-shape.

flammules — Flame-shaped markings.

fusiform — Spindle-shaped; tapering at each end.

gastropod — Molluscs of the class Gastropoda (snails and slugs).

gastrozooids — The nutritive polyps of cnidarians.

girdle — Muscular, spicule-bearing peripheral part of the mantle of a chiton, encircling the body and shell plates.

gonozooid — A sexual zooid or medusa bud of a hydroid.

helix — A spiraling coil, as in a gastropod shell.

hyaline — Transparent or semi-transparent, with a glossy or gelatinous consistency.

hydrocoral — A compound hydrozoan having a well-developed calcareous skeleton: orders Milleporina and Stylasterina.

hydroids — The polyp form of a hydrozoan (as distinguished from the medusa form); usually sessile.

infauna — Animals living in soft benthic substrats, such as sand and mud.

interambulacra — Areas of the echinoderm body that lie between rows of tube feet.

intertidal — That part of the shore zone that is alternately covered and uncovered by the tides.

iridocytes — Cells that occur in the skin of certain vertebrates and invertebrates that appear iridescent greenish or bluish from guanine.

lamella — A small, thin disc or plate.

lecithotrophic — Larvae with large yolk supply for nutrition.

lunule — An oval or crescent-shaped body part.

maculations — Spots and similar markings on an animal or plant.

mantle — The soft external body wall that covers the body of molluscs, tunicates, or barnacles.

margin — Edge or border.

marislin — A 20-carbon chemical secreted by *Chromodoris marislae* as a defense mechanism.

mesenteries — The radial muscular partitions extending inward from the wall of the digestive cavity of actinozoans.

metabolite — A metabolic by-product.

molluscivorous — Organisms that prey exclusively or primarily on molluscs.

morphology — The study of form and structure of organisms.

navenone — An intraspecific alarm substance used for chemical communication among individuals of *Navanax inermis*.

nematocysts — The stinging cells of cnidarians.

neurotoxin — A toxic protein compound in certain venomous animals that affects the nervous system.

nomenclature — A system of names used in a particular science and sanctioned by the usage of its practitioners.

notal — Of or belonging to the back.

notosetae — The chitinous bristles associated with the upper lobe of polychaete segmental locomotory appendages (parapodia).

norrisolide — A diterpene chemical secreted by the nudibranch *Chromodoris norrisi*.

ocellus — A simple eye or eyespot found in many organisms (pl. ocelli); an eye-like pigment spot on certain invertebrates, such as octopuses.

operculum — A hard or horny plug that closes the aperture of some snails, barnacles, or bryozoans.

opisthobranch — Sluglike molluscs of the order Opisthobranchia.

oral disc — The unattached end of a cnidarian polyp; in the center is a single opening to the gut cavity.

oral tentacles — Flexible appendages around the mouth.

oral veil — Quadrangular flap of skin projecting anteriorly above the mouth opening.

oscula — The large openings in sponges where water leaves the body.

ovulid — A small cowry-like gastropod of the family Ovulidae.

papilla — A small protuberance or nipple.

parapodia — Short unsegmented processes on the sides of most polychaete worms; may bear bristles and be used for locomotion.

paxillae — Spiny projections with expanded tops on the upper (aboral) body surface of some sea stars.

pearlfish — A small fish of the family Carapidae that lives commensally in the mantle cavity of the pearl oyster and also in the intestines of some sea cucumbers.

pedal disc — The attachment end of a cnidarian polyp.

pedicellaria — Specialized pincerlike appendages on the body surfaces of sea stars and sea urchins.

pelagic — Oceanic or living in the open sea.

pericardium — Region around the heart.

periostracum — A chitinous layer covering the exterior surface of many mollusc shells.

perisarc — Ectoderm-secreted chitinous covering.

petaloids — Petal-shaped ambulacral area radiating from the center of the upper body surface of some echinoderms.

phototactic — Response by an organism towards a light source.

photosynthesis — A process used by plants to convert light energy to organic compounds by combining carbon dioxide and water, catalyzed by chlorophyll, with free oxygen as a by-product.

piscivorous — Animals that prey on fishes.

planktonic — Pertaining to organisms (usually small) that drift at the mercy of water currents because they are unable to swim or are weak swimmers.

polychaetes — Annelid worms of the class Polychaeta.

polymorphism — Organisms that have, or occur in, several distinct forms.

posterior — Situated at the back end of the body, opposite the head.

prosobranch — A gastropod of the subclass Prosobranchiata.

pustulate — Covered with many small protuberances.

radula — A rasp-like organ or lingual ribbon armed with rows of teeth, found in most molluscs.

relict — Surviving as a remnant of a vanishing race, type, or species.

reno-pericardium — Blood processing (cleaning and pumping) region of kidney and heart.

reticulations — Cross-ridged, distinct lines crossing each other like a network; cancellated.

rhinophore — Sensory antenna on the head or anterior region of an opisthobranch.

rostrum — A spine-like process projecting anteriorly off the head of a crustacean.

rugose — Having a wrinkled, ridged, or roughly veined surface.

sabellid — Polychaete worms of the family Sabellidae, commonly known as feather duster worms.

sednolide — Tetracyclic terpene compound secreted by *Glossodoris sedna* for defense.

septa — The radial calcareous plates projecting into the cavity of a coral containing the polyp.

serpulid — Polychaete worms of the family Serpulidae, commonly known as tube worms.

setae — Bristles or hairs.

sipunculans — A group of marine worms of the phylum Sipuncula.

speciose — An area or taxonomic group containing a particularly large number of species.

spicules — The numerous, minute calcareous or siliceous structures supporting the tissues of various invertebrates, particularly sponges, compound ascidians, and many radiolarians.

spire — The last whorl above the body whorl of shelled gastropods.

subspecies — A taxonomic category ranking below a species whose members are morphologically distinguishable but interbreed successfully with members of other subspecies of the same species where their ranges overlap.

subterminal — Area below the ending of an extension or process; not the tip, but just below the tip.
subtidal — The region below the lowest low tide mark.
supraocular — Situated above the eye.
suspension feeding — To feed on material suspended in water such as plankton.
tactile — The sense of touch.
telson — The terminal segment of the body of an arthropod, often forming the middle lobe of the tail fan.
tentacles — Slender, flexible processes or filaments used as sense organs or for feeding and grasping.
terminal — Situated at the end or extremity of the body or appendage.
test — The fused body plates of echinoids and barnacles.
tetracyclic terpene compound — Organic 4-ringed compound of 26 carbons; a category of probably defensive chemicals isolated from nudibranchs.
toxopneustid — Sea urchins of the genus *Toxopneustes*.
taxonomy — The science of classification of living organisms.
transverse — Running across the main axis.
truncate — Square or broad at the end, as if cut off transversely.
tumid — Swollen or inflated.
tunicates — Any of the chordates of the subphylum Urochordata.
ubiquitous — Being omnipresent or everywhere, particularly at the same time.
uropods — Abdominal appendages of arthropods located on either side of the telson.
valve — One of the movable articulated plates of chitons or barnacles, or one of the two halves of clam shells.
varix — A prominent ridge across each shell whorl in many shelled gastropods showing a former position of the outer lip of the aperture (pl. varices).
veil — A flap of skin projecting from the body of an organism.
velar — Pertaining to the annular membrane projecting inward from the margin of the umbrella in hydromedusans and a few jelly fishes.
veliger — A larval stage in molluscs.
vermivorous — An animal that preys exclusively on worms.
verrucae — Warty skin tubercles.
vesicles — Animal or plant structures having the general form of a membranous cavity.
whorl — Any complete coil of a gastropod shell.
zoanthid — Any cnidarian of the order Zoanthidea; usually small colonial polyps.
zooxanthellae — Any of the various symbiotic unicelled algae that live in the tissues of other organisms such as coral polyps, giant clams, some radiolarians, and flatworms.

PHYLUM PORIFERA
Class Demospongiae (Sponges)

1. ***Aplysina fistularis* (formerly *Verongia aurea*)**
Sulfur sponge

Description: Color is variable, from bright yellow, ochre, to brown. Encrustations also vary in form from flat, coarse mats to irregular, fingerlike protuberances. *Size:* Thickness 3-6 in. (75-150 mm); diameter of encrustation 6-10 in. (150-250 mm). *Habitat:* On intertidal rocks and rocky reefs; to below 100 ft. (30.5 m). *Distribution:* Throughout the Gulf and along the outer coast of Baja California Sur and north into southern California; cosmopolitan. *Remarks:* The skeleton of this relatively common sponge, unlike most other marine sponges, is composed entirely of spongin fibers.

Tylodina fungina, a limpet-like opisthobranch, feeds on this sponge. *Verongia aurea* is now considered to be an ecophenotype of *Aplysina fistularis.*

Peggy Boyer

2. ***Tethya aurantia***

Golden spongeball

Description: Hemispherical; surface covered with evenly spaced, wart-like protuberances. Color is usually pale yellow, but may vary from orange to red. *Size:* Diameter 2-4 in. (50-100 mm). *Habitat:* Caves, crevices, and under rocks and rocky ledges; intertidal and subtidal to depths greater than 1,200 ft. (366 m). *Distribution:* Throughout the Gulf, cosmopolitan, including Europe and New Zealand. *Remarks:* This sponge is attached to hard surfaces by rootlike processes that grow out from its base.

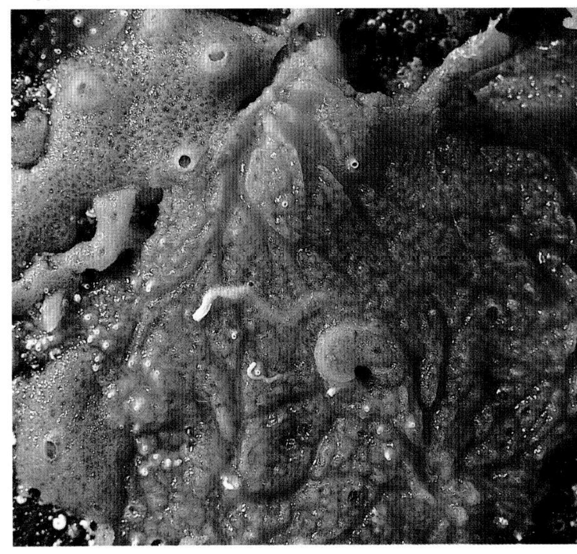

3. ***Ophlitaspongia pennata***

Furrowed sponge

Description: Color is variable, from dull yellow or light amber, to orange. Oscula are star shaped; surface is grooved with furrows. *Size:* Thickness 0.2-0.4 in. (4-10 mm). *Habitat:* Encrusting rock surfaces in the mid-and-low intertidal zones. *Distribution:* Throughout the Gulf, Canada to southern California. *Remarks:* In the Gulf, this sponge is most abundant in the mid-intertidal zone.

4. ***Hymeniacidon* sp. (species undetermined)**

Orange crust sponge

Description: Color is variable, from red-orange or pale yellow, to cream. Forms flat encrustations. Numerous oscula are randomly spaced, occasionally on raised papillae. *Size:* Thickness 2-4 in. (50-100 mm); width of encrustation 4-8 in. (100-200 mm). *Habitat:* On rocks, stones, and empty shells in the intertidal and on shallow subtidal regions. *Distribution:* Throughout the Gulf (distribution outside the Gulf is unrecorded). *Remarks:* At least three species of *Hymeniacidon* occur in the Gulf.

5. ***Acarnus erithacus***

Red velvet sponge

Description: Color is deep red to orange. Encrustations are usually flat, but are occasionally well-formed and massive. *Size:* Encrusting width 3-4 in. (75-100 mm). *Habitat:* In protected rocky crevices, under rocks and ledges; intertidal and subtidal to depths below 500 ft. (152 m). *Distribution:* Throughout the Gulf, and from west Baja California, north to British Columbia.

6. ***Pseudosuberites pseudos***

Barrel sponge

Description: Color is variable, from pinkish-brown to orange and yellow. Its cortex is typically covered with distinct tubercles. Simple encrustations occur intertidally; well-formed, often bulbous and massive encrustations occur subtidally. *Size:* Diameter of encrustation 4 in.-3 ft. (100 mm-1 m). *Habitat:* Encrustations grow on rocks and hard substrates under sand; intertidal to depths of 100 ft. (30.5 m). *Distribution:* Throughout the Gulf (distribution outside the Gulf is unrecorded). *Remarks:* This common species reaches maximum development around cliffs and offshore islands, where plankton rich currents occur.

7. ***Tedania nigrescens***

Panamic red sponge

Description: Color is normally brilliant red to orange, with a smooth surface. Oscula are occasionally difficult to see. *Size:* Width of encrustation 3-6 in. (75-150 mm) to as much as several feet; thickness to several centimeters. *Habitat:* Encrusting rock reef surfaces, rocks, or wood pilings; intertidal and subtidal. *Distribution:* Throughout the Gulf to Ecuador; cosmopolitan.

8. ***Leucetta losangelensis***

Noodle sponge

Description: Color is usually cream-white to beige. There are numerous oscula projecting at the end of "noodle-like" protuberances. *Size:* Diameter of encrustation 4-6 in. (100-150 mm); oscula diameters 0.25-0.50 in. (6-12 mm). *Habitat:* Between rocks, on dock pilings, and on boat hulls; intertidal and subtidal to below 300 ft. (91.4 m). *Distribution:* Throughout the Gulf and from west Baja California to southern California. *Remarks:* This is one of the most abundant sponges of the Gulf. A number of invertebrates live inside this sponge, including pistol shrimps, annelid worms, isopods, and amphipods.

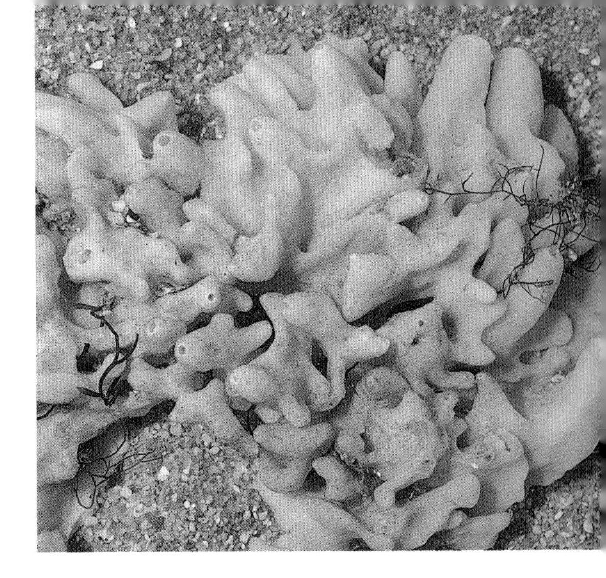

PHYLUM CNIDARIA
Class Hydrozoa (Hydroids)

9. ***Tubularia crocea***

Tubularia

Description: These hydroids may be either solitary or colonial; the stems are covered with firm covering. The tentacles are long and thin, usually lacking anchoring filaments. *Size:* Height 0.50-1 in. (12-25 mm). *Habitat:* Attached to rocky substrates, pilings, and floats in bays or harbors in shallow intertidal and subtidal regions. *Distribution:* Throughout the Gulf, around the tip of Baja California north to Alaska; also eastern and western Atlantic. *Remarks:* Species of this genus are preyed upon by various species of nudibranchs.

10. ***Corymorpha palma***

Solitary hydroid

Description: These large solitary hydroids have reduced exoskeleton and long, thin tentacles that often have a knob on the end. There are numerous fine anchoring filaments at the base of the stalks. *Size:* Polyp height to 6 in. (150 mm). *Habitat:* Mudflats of the low intertidal zone, in bays, and protected areas. *Distribution:* Throughout the Gulf, also southern California. *Remarks:* This locally abundant species exhibits a complex behavior, remaining upright for 3-8 minutes and then bending over until the polyp's tentacles touch the substratum. By sweeping its tentacles over the sediment, it picks up food particles.

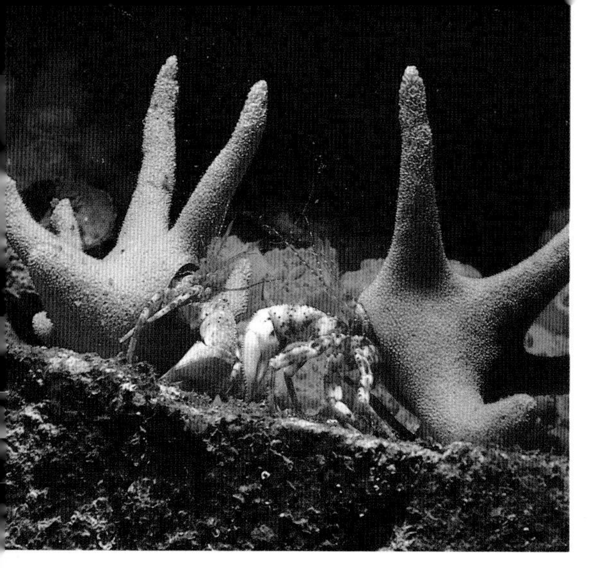

Ron McPeak

Hans Bertsch

11. ***Janaria mirabilis***

Staghorn hydrocoral

Description: This hydrocoral typically assumes a helmet shape, with crown-like central branches and a basal lateral branch. The branches are more or less symmetrical. *Size:* Height 1-2 in. (25-50 mm). *Habitat:* Sand or rubble substrates, to depths of 10-500 ft. (3-152 m). *Distribution:* Central and southern Gulf (distribution outside the Gulf is unrecorded). *Remarks:* These colonial hydroids encrust and eventually dissolve gastropod shells that house the hermit crab, *Manucomplanus varians* (formerly *Pylopagurus varians*). The hermit crab trims and manicures the entrance-way as the hydrocoral continually grows. The bizarre pronged growth of the hydrocoral prohibits the hermit crab from easily righting its house if overturned. This coral is fairly common in some localities.

12. ***Aglaophenia diegensis***

Ostrich plume hydroids

Description: The clumps of feather-like stalks are usually brown, yellowish-brown, or greenish-brown. *Size:* Height 4-6 in. (100-150 mm). *Habitat:* Rocky substrates in both the intertidal and subtidal zones. *Distribution:* Throughout the Gulf to Ecuador. *Remarks:* The prominent, brownish, rice grain-shaped structures are corbulae, which contain reduced medusae. These medusae are not liberated as free-swimming planktonic organisms, but are confined within the corbulae from where they either release sperm or produce eggs.

13. ***Lytocarpus nuttingi***

Stinging hydroid

Description: The feather-like branches have a silvery sheen. *Size:* Clump diameter 4-12 in. (100-300 mm). *Habitat:* Shallow rocky subtidal areas, to about 65 ft. (20 m). *Distribution:* Central and southern Gulf, west Baja California to Bahía Magdalena (distribution outside the Gulf is unrecorded). *Remarks:* The cryptic nudibranch, *Lomanotus stauberi,* lives on, feeds on, and lays its egg masses on *Lytocarpus*. Although the sting of this hydroid is painful, it is usually not lasting and is considered only a nuisance by most divers.

14. ***Physalia utrilculus***

Pacific man-of-war

Description: The bluish-purple, gas filled, oval float, below which dangle numerous specialized polyps, is attached to the highly muscular and contractile stinging tentacles. *Size:* Swimming float length 2-3 in. (50-75 mm) in the Gulf; to 12 in. (300 mm) elsewhere. *Habitat:* Pelagic; floating on the surface of the water, carried about by winds and currents. *Distribution:* Throughout the Gulf to Panama, and the Indo-Pacific. *Remarks:* The tentacles bear large stinging cells which can inject a potent neurotoxin. Human reactions to stings vary from transient local pain to severe reactions requiring hospitalization. The commensal fish, *Nomeus gronovii,* lives among *Physalia*'s stinging tentacles.

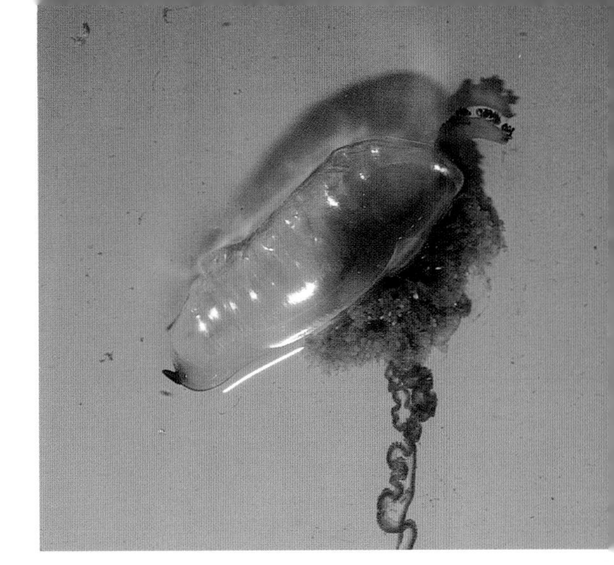

15. ***Porpita pacifica***

Porpita

Description: The small, flattened, disc-shaped animal does not have a conspicuous gas-filled float. *Size:* Diameter 0.5-2 in. (12-50 mm). *Habitat:* Pelagic; floating on the surface of the water. *Distribution:* Throughout the Gulf; circumtropical and subtropical. *Remarks:* Attached below the flat upper disc of the colony are three types of tentacles: the outermost dactylozooids (short, hollow, nematocyst-bearing offensive and defensive members of the colony); a central gasterozooid (with true mouth); and an intermediate band of gonozooids from which gamete-bearing medusae are set free. They are often blown ashore in large numbers by storms and onshore winds.

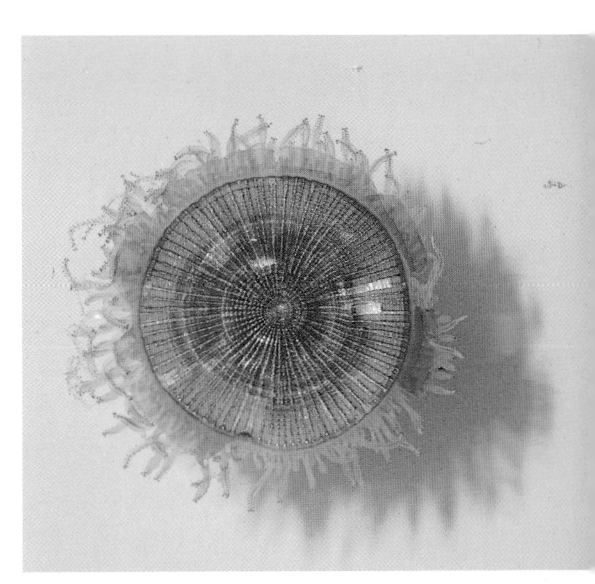

16. ***Apolemia uvaria***

Fringed stem medusa

Description: The undulating, lace-like colony has a contractile stem and a very small float. *Size:* Colony length 15-25 ft. (5-8 m). *Habitat:* Floating in mid-water and near the surface to below 100 ft. (30.5 m). *Distribution:* Throughout the Gulf to Peru; circumtropical. *Remarks:* Colony is made up of several hundred individuals. As in the man-of-war, the tentacles possess stinging cells capable of delivering painful stings to divers and bathers. They are seasonally common in some areas.

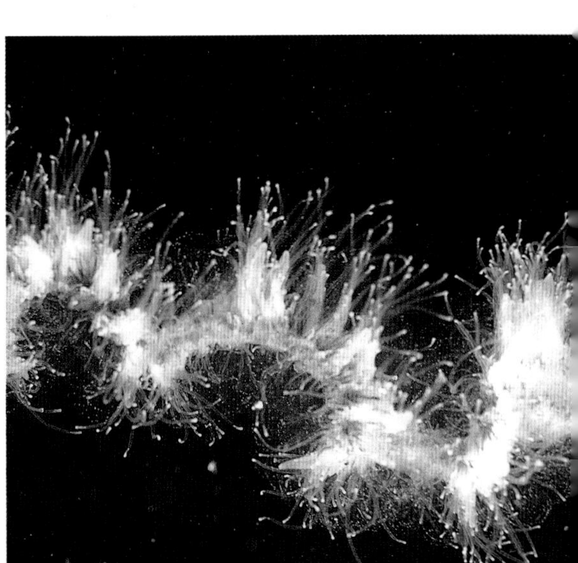

Peter Pickens

Class Anthozoa (Anemones, Corals, Gorgonians)

17. *Phialoba steinbecki*

Steinbeck's phialoba anemone

Description: The most distinctive feature is the lobed oral disc with numerous short tentacles. More than 400 tentacles may be present. The tentacles are green or olive green, sometimes pink-tipped. The base color is green, olive green, or whitish; white stripes often occur on the oral disc. *Size:* Oral disc diameter 1.5-3 in. (38-75 mm); column height 1 in. (25 mm). *Habitat:* On rock surfaces, in crevices, or at the interface between rock and sand; intertidal to at least 20 ft. (6.5 m). *Distribution:* Throughout the Gulf (distribution outside the Gulf is unrecorded).

18. ***Bunodosoma californica***

Teflon anemone

Description: Column covered with closely-packed, non-adhesive warts called verrucae. Verrucae are arranged in longitudinal rows only at the base; elsewhere they are randomly placed over the column. Column is red, brown, or olive green. There may be yellow hashmarks around the mouth. Individuals from the central and southern Gulf may lack the light hashmarks around the mouth, but may have white patches on the tentacles. *Size:* Oral disc diameter 1.5-3 in. (38-75 mm); column height 3 in. (75 mm). *Habitat:* On intertidal and subtidal rocks and reefs. *Distribution:* Throughout the Gulf, west coast of Baja California to Bahía Magdalena.

19. ***Bunodactis mexicana***

Sticky wart anemone

Description: Column covered with warts (verrucae) arranged in tightly packed longitudinal rows. Column is similar in color to *Bunodosoma* (red, brown, or olive green). Tentacles are usually as long as column, olive green, and often with a white fusiform mark around the base of each inner tentacle but without obvious white patches on tentacles. Tentacles may number over 100 and are often reddish-tipped. *Size:* Oral disc diameter 2.5

in. (64 mm); column height 2-2.5 in. (50-64 mm). *Habitat:* On intertidal and shallow subtidal rocks. *Distribution:* Throughout the Gulf (distribution outside the Gulf is unrecorded).

20. ***Anthopleura dowii***

Dow's anthopleura

Description: The warts are arranged in widely-spaced rows showing exposed areas of the column between them. It resembles *Bunodactis* but is smaller and differentiated by the organization of verrucae. The rows of verrucae are fewer, usually larger at the top and straighter than *Bunodactis*. *Anthopleura* is dark green or brown and has transparent tentacles with white patches. *Size:* Column length 1-2 in. (25-50 mm); tentacle length 0.5-1 in. (12-25 mm). *Habitat:* Intertidal and subtidal rocks and reefs. *Distribution:* Throughout the Gulf to Ecuador.

21. ***Alicia beebei***

Sand anemone

Description: During the day this anemone retracts into a flat, bubbly, jelly-like mass. When expanded at night, scattered compound outgrowths are apparent on the column. These outgrowths are large and branched. Tentacles and the upper part of the column have white spots. The contracted anemone is completely covered with branching outgrowths. *Size:* Expanded column length 4-8 in. (100-200 mm); tentacle length 8-15 in. (200-375 mm). *Habitat:* Usually under intertidal and subtidal rocks to depths of at least 82 ft. (25 m). *Distribution:* Central Gulf to Panama. *Remarks:* During the day it is usually hidden in crevices or under rocks. It comes out at night and resembles the tube anemone, *Pachycerianthus*.

22. ***Telmatactis panamensis***

Panamanian clubbed anemone

Description: The short, club-shaped tentacles may number to 100. The column and tentacles vary in color from orange, red, and rose to brown. The anemone has acontia, and the column is covered with a parchment-like material. *Size:* Oral disc diameter 1-2 in. (25-50 mm); column height 3-4 in. (75-100 mm). *Habitat:* It occurs deep in crevices or beneath rocks, subtidal to depths of 10-200 ft. (3-61 m). *Distribution:* Throughout the Gulf to Panama. *Remarks:* Often this species shares the same cre-

vice as the scallop, *Lima tetrica*. It is usually contracted during the day, but expands at night to feed.

23. ***Calliactis polypus* (formerly *C. variegata*)**
Hitch-hiker anemone
Description: The column color is light brown with variable splotches or lines of red. There are white or light colored warty structures near the base. The tentacles are light, mottled brown, and number over 200. *Size:* Oral disc diameter 2 in. (50 mm); column height 2 in. (50 mm). *Habitat:* This anemone is usually found attached to shells occupied by the hermit crab, *Dardanus sinistripes;* intertidal and subtidal. *Distribution:* Throughout the Gulf to Panama. *Remarks:* The anemone-hermit crab relationship involves complex behavioral interactions. A hermit crab can transfer anemones from discarded shells to new shells rapidly by lightly stroking the column which induces the anemone to detach itself from the old shell.

24. ***Tealia piscivora***
Fish-eating anemone
Description: The most distinctive characteristics of this species are the large size and smooth red to maroon column. The tentacles are pink with diamond patches at the base or pure white with yellow patches at the base, at least on the inner most tentacles. The oral disc has a large tentacle-free area around the mouth. *Size:* Oral disc diameter 12-14 in. (30-35 cm); column height 10 in. (25 cm). *Habitat:* Rocky substrates, at depths below 165 ft. (50 m). *Distribution:* This animal was found off Shepard's Rock, Cabo San Lucas, at a depth of 180 ft. (55 m). *Tealia piscivora* has been reported previously from southern California to southern Alaska. *Remarks:* This is the first report of *Tealia piscivora* from Baja California Sur.

25. ***Tealia columbiana***
Sand rose anemone
Description: The most distinctive features of this large anemone are the white or cream-colored verrucae on the column. The column is red or maroon, the oral disc maroon, and the tentacles white or cream. *Size:* Basal diameter 14 in. (35 cm); column height 10 in. (25 cm). *Habitat:* Rocky substrates usually covered by sand or calcareous debris at depths greater than 165 ft. (50 m). *Distribution:*

This animal was found off Shepard's Rock, Cabo San Lucas, at a depth of 180 ft. (55 m). *Tealia columbiana* had been reported previously from southern California to British Columbia, Canada. *Remarks:* This is a new distributional record for a cooler water species.

26. ***Aiptasia californica***

Transparent anemone

Description: The overall color is usually a dull brown, greenish brown, or translucent white. When the column is well extended, the acontia, filaments, and actinopharynx are visible through the column. *Size:* Column height 2 in. (50 mm); tentacle length 2 in. (50 cm). *Habitat:* Covers rocky edges of tidepools in low intertidal zones. *Distribution:* Throughout the Gulf (outside the Gulf it has only been recorded from San Diego, California). *Remarks:* This prolific anemone reproduces asexually by pedal laceration. Symbiotic algae, called zooxanthellae, live in the anemone's tissues and are responsible for its color.

27. ***Phyllactis bradleyi***

Collared anemone

Description: This genus is easily distinguished from all other Gulf anemones by the presence of a frill or collar below the tentacles. The collar bears many short filiform or dendritic appendages. The tentacles are frequently retracted and only the collar and oral disc show on the surface. Tentacles are set close to the mouth and when fully extended can be quite long. *Size:* Oral disc and collar diameter 2 in. (50 mm); column height 5 in. (125 mm). *Habitat:* Lowest intertidal and subtidal sand or mud substrates. Although most of the anemone is usually buried in sand, the oral surface and tentacles can be found exposed just above the sand. *Distribution:* Throughout the Gulf to Panama; also in southern California. *Remarks:* This common species is similar to another Gulf collared anemone, *Phyllactis cocinnata,* the two may be ecotypes of the same species.

28. ***Antiparactis* sp. (species undetermined)**

Gorgonian wrapper

Description: The column is light orange with scattered brown spots. There are approximately 100 light orange tentacles. *Size:* Oral disc diameter 2 in. (50 mm); column height 1.5-2 in. (38-50 mm). *Habitat:* Attaches to gorgonians; in shallow subti-

Ron McPeak

dal depths, to at least 100 ft. (30.5 m). *Distribution:* Throughout the Gulf to central Mexico, including Islas Revillagigedos. *Remarks:* The gorgonian wrapper attaches itself to various gorgonians, including *Lophogorgia alba, Eugorgia aurantica,* and *Muricea* sp. This anemone is probably an undescribed species.

29. ***Epizoanthus* sp. (species undetermined)**
Red epizoanthid

Description: The column and tentacles are red to reddish-brown in color. The column epidermis is usually covered with sand or mud. *Size:* Polyp height 1 in. (25 mm). *Habitat:* On rocky reefs, sand, and rubble substrates, encrusting gorgonian stalks or covering rock surfaces; shallow subtidal to at least 250 ft. (76.2 m). *Distribution:* Throughout the Gulf. *Remarks:* This species is particularly common in the upper and central Gulf. An undescribed genus and species of nudibranch lives and feeds on this zoanthid.

30. ***Palythoa ignota***
Colonial zoanthid anemone

Description: The polyp color is brown to olive-green. Like other zoanthid anemones, this species has only one ciliated groove on the oral disc and lacks a pedal disc. The column epidermis is relatively thick. *Size:* Polyp diameter 0.5-1 in. (12-25 mm). *Habitat:* Covers rocks and reef surfaces, particularly along tide pool edges; low intertidal and shallow subtidal. *Distribution:* Northern and central Gulf (distribution below the central Gulf is unrecorded). *Remarks:* This is one of the most abundant zoanthid anemones within its range. Its coloration results from zooxanthellae, symbiotic one-celled algae living in its tissues.

31. ***Pachycerianthus fimbriatus***
Pacific tube anemone

Description: The long, thin tentacles extending out of the mucus-like tube vary in color from white to light blue, purple or pink and are occasionally bi-colored. *Size:* Total expanded length 10-15 in. (250-375 mm). *Habitat:* Partially buried under sandy or rubble substrates; depths of 30-150 ft. (10-46 m). *Distribution:* Throughout the Gulf, northern to southern California (distribution south of the Gulf is poorly known). *Remarks:* This elegant species is most often encountered with its numerous tentacles exposed above the sand or rubble. When disturbed it quickly withdraws into its tube with a rapid jerk. The cleaner shrimp, *Periclimenes lucasi,* is often associated with this anemone, living at the base of the tube.

32. ***Pachycerianthus insignis***

Whitespotted tube anemone

Description: The thin, long tentacles are variable in color, from brown, to purple or light pink, and bear white spots that occasionally coalesce into stripes running the length of each tentacle. *Size:* Column length 2-4 in. (50-100 mm); tentacle length 3-4 in. (75-100 mm). *Habitat:* Buried in sand, mud or rubble; intertidal and subtidal to depths of 100 ft. (30.5 m). *Distribution:* Throughout the Gulf to Panama. *Remarks:* When exposed, the spread of the tentacles remains relatively flat and close to the sand, rather than elevated as in most cerianthids.

33. ***Pocillopora elegans***

Elegant coral

Description: The branching growths produce large, bushlike coral heads which range in color from pale to dark brown to green. *Size:* Head height 2-3.5 ft. (0.75-1 m) *Habitat:* Shallow subtidal, at depths of 3-50 ft. (1-17 m). *Distribution:* Baja side of the central Gulf, from Isla Carmen to Cabo San Lucas, and south to Ecuador. *Remarks:* *Pocillopora* forms the only true "coral reefs" in the Gulf, near Cabo Pulmo and Los Frailes. The cavities in the coral heads provide an excellent habitat for the coral hawkfish, *Cirrhitichthys oxycephalus,* and a number of invertebrates.

Ron McPeak

34. ***Pavona gigantea***

Giant coral

Description: The polyps are grey, with short tentacles. The coral heads usually form non-branching, flat, or round-topped columnar growths. *Size:* Encrusting colonies 3-8 ft. (1-2.75 m) in area. *Habitat:* Shallow subtidal, rarely below 30 ft. (10 m). *Distribution:* Baja side of the central Gulf, from Isla Carmen to Cabo San Lucas, and south to Ecuador. *Remarks:* This species is especially common on the reefs at Cabo Pulmo.

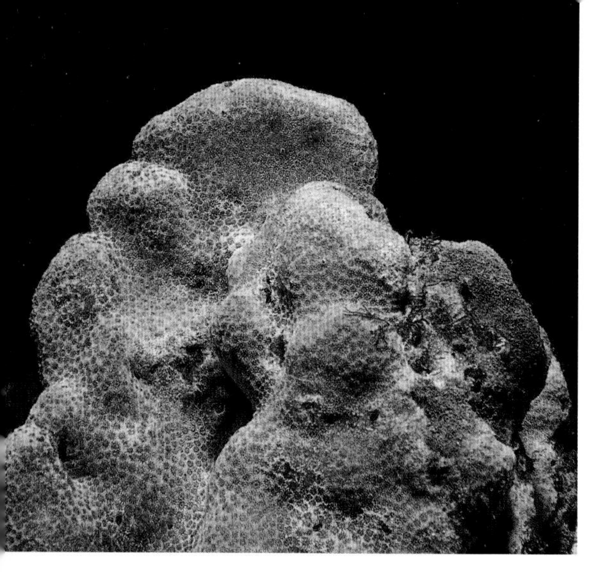

35. ***Porites californica***

Emerald coral

Description: Living colonies are emerald green in shallow water to yellowish-brown in deeper water. Colonies vary in form, from flat encrustations to tall, branching forms. Well-developed, branching encrustations are relatively rare.. *Size:* Thickness of encrusting colonies 3-5 in. (75-130 mm); covering area from 1-20 in. (25-500 mm). *Habitat:* Commonly encrusting intertidal and subtidal rocks; rarely below 60 ft. (20 m). *Distribution:* Throughout the Gulf to Panama. *Remarks:* The polyp tissues contain unicellular, chlorophyll-bearing photosynthetuc zooxanthellae, which are responsible for the green color of shallow water specimens. The crown-of-thorn sea stars, *Acanthaster ellisii,* prey on this coral's polyps.

36. ***Astrangia cortezi***

Cortez coral

Description: Color is generally orange, light orange, or pink. The corallites of these delicately branching, arborescent colonies are trumpet-shaped, wider distally, and narrower at the base. *Size:* Colony diameter 6-15 in. (150-375 mm); corallite diameter 0.1-0.2 in. (3.5-5.5 mm). *Habitat:* Beneath rocky overhangs or in shallow caves, at depths of 15-115 ft. (5-35 m). *Distribution:* Throughout the Gulf (distribution outside the Gulf is unknown). *Remarks:* This coral is locally common, particularly along offshore islands where currents are swift. It was originally described from specimens collected at Consag Rock in the northern Gulf.

Ron McPeak

37. ***Coenocyathus bowersi***

Bowers' cup coral

Description: Colony consists of tight clusters of light orange to pink corallites. *Size:* Colony diameter 4-5 in. (100-125 mm); corallite diameter 0.12-0.5 in. (3-12 mm). *Habitat:* In caves or reef overhangs; subtidal at depths of 30-490 ft. (10-150 m). *Distribution:* Central and southern Gulf, and Monterey (California) to southern California (distribution south of the Gulf is unrecorded). *Remarks:* Locally common throughout its known range, particularly where currents are swift.

38. ***Bathycyathus consagensis***

Consag cup coral

Description: There are six or seven calcareous radial plates that extend above the surface of the cup. The body color is brown to reddish-brown; the texture of contracted polyps appears swollen and fleshy, with transparent tentacles. *Size:* Corallite diameter 0.25-0.37 in. (6-9 mm). *Habitat:* On rocky substrates, often beneath overhangs to depths of 325 ft. (100 m). *Distribution:* Northern and central Gulf and outer Baja California from Bahía Magdalena to southern California.

Ron McPeak

39. ***Tubastraea coccinea* (formerly *Tubastraea tenuilamellosa*)**

Orange cup coral

Description: Living colonies are bright yellow to orange, with corallites that are separate from one another but fused at the base. *Size:* Clump diameter 1-12 in. (25-300 mm); corallite diameter 0.5-1 in. (12-25 mm). *Habitat:* Colonies are especially common in shaded areas, beneath overhangs and under ledges or even under large boulders, from the lowest intertidal to at least 60 ft. (20 m). *Distribution:* Central and southern Gulf to Ecuador; circumtropical. *Remarks:* Colonies of *Tubastraea* serve as both food and habitat for the gastropod *Epitonium billeeanum*. Common on off-shore islands of the central Gulf, becoming more abundant further south.

40. ***Antipathes galapagensis***

Yellow polyp black coral

Description: The branching colony consists of a spiny skeleton bearing bright yellow, nonretractable polyps. *Size:* Colony diameter 1-6 ft. (0.3-2 m). *Habitat:* Anchored to rocky reefs at depths of 10 to at least 250 ft. (3-76 m). *Distribution:* Central and southern Gulf to Ecuador and the Galápagos. *Remarks:* This locally common coral is particularly abundant below 150 ft. The shrimp, *Periclimenes infraspinis,* and the longnose hawkfish, *Oxycirrhites typus,* live among the coral's branches. This species is currently harvested commercially for jewelry in La Paz and Cabo San Lucas. In shallow water, the branching colonies are small, scraggly, and poorly developed.

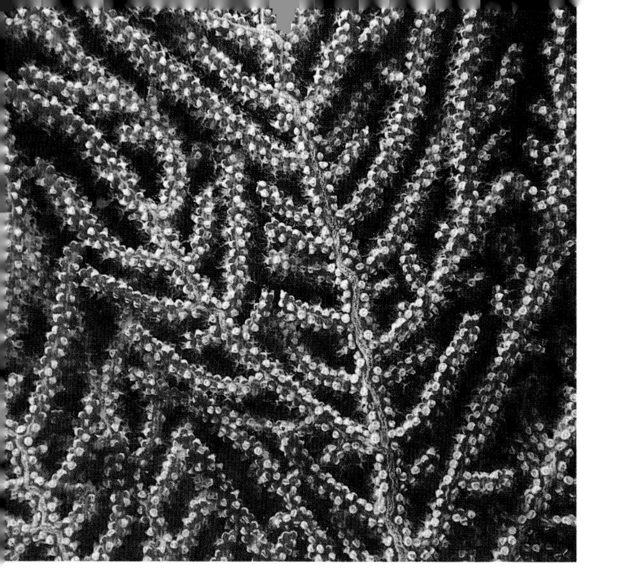

41. ***Eugorgia aurantica***

Bumpy orange gorgonian

Description: Red and yellow branches with yellow-white polyps. *Size:* Height 25-30 in. (60-80 m). *Habitat:* Rocky subtidal. *Distribution:* Throughout the Gulf and outer coast of Baja California from Cabo San Lucas to Isla Cedros. *Remarks:* The gorgonian wrapper anemone, *Antiparactis,* often attaches itself to this gorgonian.

Hans Bertsch

42. ***Lophogorgia alba***

White gorgonian

Description: The long, slender white stalks taper to pointed ends and bear white polyps. *Size:* Height 12-35 in. (30-90 cm). *Habitat:* Rocky reefs; shallow subtidal to 165 ft. (50 m). *Distribution:* Throughout the Gulf to Panama. *Remarks:* A white shell, *Simnia aequalis,* covered with a dark-spotted, white mantle, can often be found on this species. It is locally common, particularly in certain areas of Bahía de los Angeles.

Hans Bertsch

43. ***Pacifigorgia* sp. (species undetermined)**

Panamic sea fan

Description: Branches grow in one plane (or two, at 90° to each other), and interlock to form a flat meshlike network. The "fan" shape is unmistakable. Overall color of branches is variable, from rusty-red or brown with yellowish polyps, to purple with white polyps. *Size:* Width 4-12 in. (100-300 mm). *Habitat:* On rocky subtidal substrates. *Distribution:* Throughout the Gulf. *Remarks:* This species, like most other sea fans, usually grows with the flat side facing the prevailing water current or surge, thus facilitating filter feeding of small particles from the water. Described (known) species from this region include *Pacifigorgia adamsii* and *P. irene.*

44. ***Muricea californica***

Robust gorgonian

Description: The color of this gorgonian varies from light yellowish-brown to reddish-brown or purple. When the long, whitish polyps are extended the branches appear thick and bushy. *Size:* Height 8-32 in. (200-800 mm). *Habitat:* Attached to rocky substrates in the extreme low intertidal and subtidal, to 180 ft. (55 m). *Distribution:* Throughout the Gulf to Panama; tip of Baja California, north to Newport Bay (southern California). *Remarks:* This gorgonian is abundant in some localities, particularly where currents are swift. The ovulid cowry, *Cyphoma emarginatum,* is occasionally found on its branches, feeding on its polyps.

45. ***Ptilosarcus undulatus***

Fleshy sea pen

Description: The flat central stalk bears fleshy lobes, with yellow-orange to vivid orange peripheral trim. *Size:* Height 8-12 in. (200-300 mm); diameter 1-4 in. (25-100 mm). *Habitat:* Muddy and sandy substrates; subtidal at depth of 33-165 ft. (10-50 m). *Distribution:* Throughout the Gulf (distribution outside the Gulf is poorly known) *Remarks:* This species closely resembles the north and central eastern Pacific species, *Ptilosarcus gurneyi.* The fleshy sea pen can withdraw rapidly into the sand when disturbed. It is never common, but fair numbers are occasionally taken by shrimp trawlers. It is preyed upon by the burrowing nudibranch, *Histiomena convolvula.*

46. ***Stylatula elongata***

Slender sea pen

Description: This delicate, thin sea pen varies in color from salmon to uniform white. *Size:* Height 4-8 in. (100-200 mm) in the Gulf, to 24 in. (600 mm) outside the Gulf. *Habitat:* Sandy or muddy substrates at depths of 15-230 ft. (5-70 m). *Distribution:* Throughout the Gulf, and from southern California to British Columbia. *Remarks:* The nudibranch, *Armina californica,* feeds on this sea pen in southern California. The nudibranch poises itself close to *Stylatula,* then suddenly lunges and takes a bite; the sea pen immediately retracts below the sand, leaving *Armina* with a one-bite meal!

Hans Bertsch

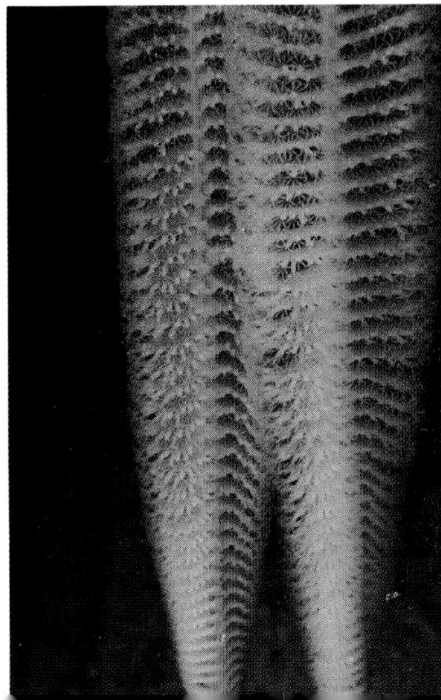

PHYLUM PLATYHELMINTHES
Class Turbellaria (Flatworms)

47. ***Thysanozoon californicum***

Mexican skirt dancer flatworm

Description: The short, dorsal, finger-like papillae distinguish this flatworm from all other Gulf species. *Size:* Body length 1-2 in. (25-50 mm). *Habitat:* Under rocks; intertidal and subtidal to at least 50 ft. (15.2 m). *Distribution:* Northern and central Gulf. *Remarks:* The dorsal papillae may be capable of releasing a strong acidic secretion when this worm is threatened by predators.

48. ***Pseudoceros bajae***

Speckled flatworm

Description: Color varies from black, tan to orange, with white spots that extend to the body margin. It has marginal tentacles on the head and a sucker on the underside. *Size:* Body length 1-2 in. (25-50 mm). *Habitat:* Under intertidal and subtidal rocks; to at least 40 ft. (12.2 m). *Distribution:* Throughout the Gulf (distribution outside the Gulf has not been recorded). *Remarks:* Unlike the related species, *Pseudoceros mexicanus,* the speckled flatworm will not readily swim when threatened by predators, but prefers to crawl to safety. This species is common throughout the Gulf.

PHYLUM NEMERTEA
Class Anopla (Ribbon Worms)

49. ***Baseodiscus mexicanus***

Zebra worm

Description: The unmistakable "zebra" pattern distinguishes this species from all other Gulf worms. *Size:* Body length 3-12 ft. (1-4 m). *Habitat:* Under rocks; intertidal and subtidal to depths of 80 ft. (24.4 m). *Distribution:* Throughout the Gulf. *Remarks:* This common species is nocturnal, emerging from under rocks at night. These worms have frequently been mistaken for sea snakes by scuba divers and snorkelers on night dives. When contracted, the body length is about ¼ the expanded length.

PHYLUM ANNELIDA
Class Polychaeta
(Segmented Worms)

50. ***Eurythoe complanata***

Common fireworm

Description: Body coloration ranges from light pink to pinkish-blue, with bushy spines. Has one pair of dorsal cirri per segment. *Size:* Body length 4-6 in. (100-150 mm). *Habitat:* Under intertidal and subtidal rocks, to depths of 200 ft. (61 m). *Distribution:* Throughout the Gulf to Panama. *Remarks:* Like many other amphinomid worms, this species can inflict painful burning stings with its venomous bristles. Although stings from fireworms are not usually serious, complications due to secondary infection have been reported.

51. ***Chloeia viridis***

Ornate fireworm

Description: The body color is variable from light iridescent greenish-white to brownish pink, with ornate feathery gills. The bristles are whitish and occasionally tipped with red. *Size:* Body length 3-5 in. (75-125 mm). *Habitat:* Under intertidal and subtidal rocks, on sand or muddy substrates, to below 300 ft. (91 m). *Distribution:* Throughout the Gulf to Panama, and the entire Caribbean. *Remarks:* The bristles of this species are venomous and will break off into the skin if touched, inflicting painful stings.

52. ***Aphrodita refulgida***

Mexican sea mouse

Description: Can be distinguished from other polychaete worms by the presence of heavy, iridescent golden spines and a flattened body covered dorsally with fine, silk-like bristles. *Size:* Body length 3-5 in. (75-125 mm). *Habitat:* Offshore, on muddy bottoms; at depths of 150-400 ft. (45.7-122 m). *Distribution:* Central and southern Gulf (distribution outside the Gulf is poorly known). *Remarks:* This common worm is often captured by shrimp trawlers in large numbers and is usually covered with muddy silt, obscuring the golden spines. Occasionally a commensal bivalve, *Neaeromya rugifera,* lives attached to the underside of this worm.

53. ***Mesochaetopterus mexicana***

Parchment worm

Description: Mid-body with large, bilobed, paddle-like parapodia. The anterior portion of the body has 14 segments. *Size:* Body length 4-6 in. (100-150 mm). *Habitat:* These worms live in parchment tubes, open at both ends, in sand or mud; intertidal and subtidal to 30 ft. (9.1 m). *Distribution:* Northern and central Gulf (may be a Gulf endemic). *Remarks:* The paddle-like parapodia draw food-rich water through a filtering mucus net. Periodically, the food-laden net is conveyed to the mouth and ingested.

54. ***Bispira rugosa monterea***

Panamic fanworm

Description: The banded branchial crown is inrolled dorsally. Color is not as variable as in many other serpulids. *Size:* Body length 1-2 in. (25-50 mm). *Habitat:* Emerging from rocks, living in membranous tubes; intertidal and subtidal to 200 ft. (61 m). *Distribution:* Throughout the Gulf to Panama. *Remarks:* The tentacles can be rapidly withdrawn into the tube when disturbed. Unlike serpulid fanworms, *Spirobranchus,* sabellid fanworms do not live in calcareous tubes, but construct mud or parchment tubes. They do not have an operculum that seals the tube when the tentacles are withdrawn.

55. ***Spirobranchus giganteus***

Spiral gilled tube worm

Description: The branchial crown coloration is highly variable, from white, blue, yellow, or red, with some individuals being multi-colored. The crown is formed by two branches of spiraling gills that may be seen emerging from a calcareous tube. The opening of the tube bears 2-4 sharp antler-like processes. *Size:* Body length 5 in. (125 mm) with 200 segments. Crown diameter (opened) 1-2 in. (25-50 mm). *Habitat:* Attached to rocks, coral, or sponges; low intertidal and subtidal to below 200 ft. (61 m). *Distribution:* Central Gulf to Panama; circumtropical. *Remarks:* The tentacles forming the branchial crown are used for respiration as well as for filtering plankton. Each worm possesses a calcareous operculum used to cover the opening of the tube when the tentacles are withdrawn.

56. ***Filograna implexa***

Coralline tubeworms

Description: Individual worms are salmon colored and have eight tiny tentacles. They live in intertwined calcareous tubes resembling coral clusters. *Size:* Length of individual worm slightly less than 0.25 in. (6 mm); colonies reach 6-12 in. in diameter (150-300 mm). *Habitat:* Attached to intertidal and subtidal rocks, to 150 ft. (45.7 m). *Distribution:* Throughout the Gulf; circumtropical. *Remarks:* Colonies are often mistaken for small coral heads. The calcareous tube clusters are very brittle and easily broken when handled.

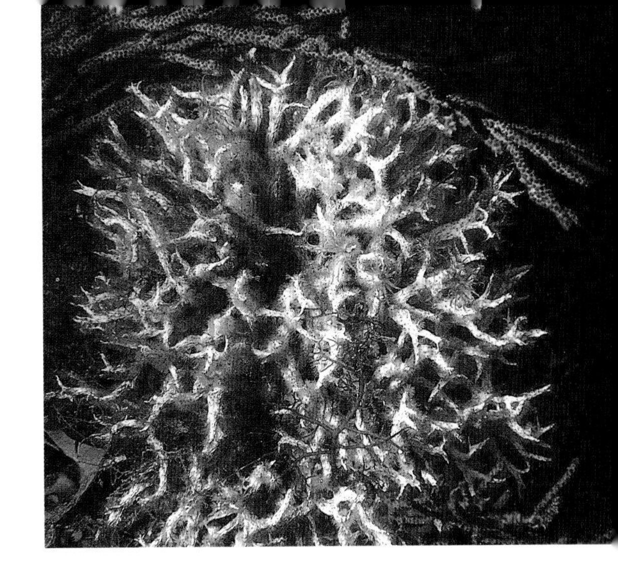

PHYLUM SIPUNCULA
Class Gephyrea (Peanut Worms)

57. ***Sipunculus nudus***

Giant peanut worm

Description: The pearly white skin is tough, with a latticed surface texture, and under proper lighting appears opalescent. *Size:* Body length 4-5 in. (100-125 mm). *Habitat:* Estuaries, mud flats, and offshore in sand or under rocks; intertidal and subtidal. *Distribution:* Throughout the Gulf, central California to Panama; cosmopolitan. *Remarks:* The mouth of sipunculans is located at the tip of a protrusible appendage (the proboscis) which can extend over 7 in. (175 mm) out of the body when searching for food.

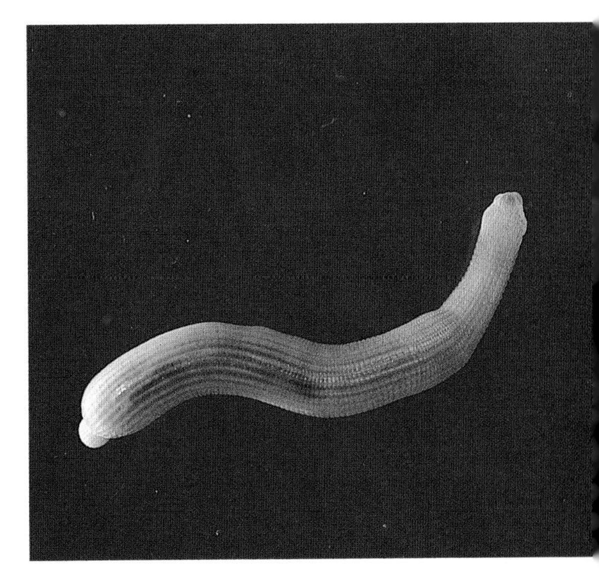

PHYLUM MOLLUSCA
Class Pelecypoda (Clams, Oysters)

58. ***Pinctada mazatlanica***

Panamic pearl oyster

Description: The outer shell valves are grayish brown, relatively thin and scaly. The nacreous interior surface is pearly greenish to brown. *Size:* Shell length 4-8 in. (100-200 mm). *Habitat:* Attached to subtidal rocks by strong byssal fibers; at depths of 10-150 ft. (3-46 m). *Distribution:* Throughout the Gulf to Peru. *Remarks:* This species was once actively harvested for its pearls in the La Paz area. Today it is gathered mostly for food. A pair of commensal shrimps, *Pontonia margarita* (male and female), are nearly always found

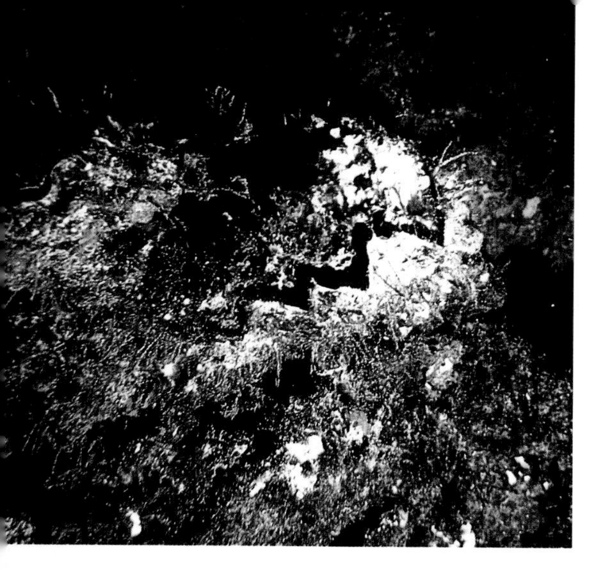

living inside the live shells. Occasionally the pearl fish, *Carapus* sp., is also associated with this species.

59. ***Hyotisa hyotis* (formerly *Ostrea fischeri*)**
Giant oyster

Description: the rough surface of the upper valve is heavily encrusted and drilled by boring animals. The inside of the valves are brown to blackish purple. *Size:* Shell length 6-10 in. (150-250 mm). *Habitat:* Attached to the underside ledges of reefs and inside caves; subtidal to depths of 10-200 ft. (3-61 m). *Distribution:* Throughout the Gulf to Ecuador; also Madagascar to the Tuamotu Islands in the Indo-West Pacific. *Remarks:* Common in some localities. Several individuals may be attached to each other. In some cases entire cave walls are covered with layers of these bivalves.

60. ***Pinna rugosa***
Panamic pen shell

Description: The triangular shell has approximately eight or nine rows of tubular spines. *Size:* Shell length 8-24 in. (200-600 mm). *Habitat:* Partially buried vertically in sand, rubble, or mud. Always anchored to rocks under the substrate by fine byssal fibers; intertidal and subtidal to depths of 100 ft. (30.5 m). *Distribution:* Throughout the Gulf to Peru. *Remarks:* A pair of commensal shirmp, *Pontonia* sp. (male and female) is nearly always found living inside each pen shell. The pearl fish, *Carapus* sp., is also occasionally found inside the living shell.

61. ***Argopecten circularis***
Pacific calico scallop

Description: Both valves are convex and sculptured with approximately 21 ribs. The shell color is extremely variable, from pure white to mottling patterns of purple, brown, yellow, or bright orange. *Size:* Shell length 1-3 in. (25-75 mm). *Habitat:* On intertidal and subtidal sandy bottoms; to a depth of 400 ft. (122 m). *Distribution:* Throughout the Gulf to Peru. *Remarks:* Like other scallops, this species is able to swim away from predators, such as sea stars, by jet propulsion, effected by rapidly opening and closing its valves. Large numbers are commonly caught by shrimp trawlers.

62. ***Lyropecten subnodosus***

Pacific lion's paw

Description: The largest of Panamic scallops. The shell is usually dull brown but occasionally varies from purple to bright orange. Although often eroded away, conspicuous, well-developed knobs on the ribs are normally located on the upper valve. *Size:* Shell length 5-8 in. (125-200 mm). *Habitat:* On subtidal rubble and sand; at depths of 10-200 ft. (3-61 m). *Distribution:* Throughout the Gulf to Peru. *Remarks:* This common scallop is harvested in enormous quantities by commercial divers off Baja California Sur.

63. ***Spondylus calcifer***

Purplelip rock oyster

Description: The exterior shell is dark reddish-brown, with a distinct purple band along the inside margin of the lip. The shell is usually heavily encrusted and drilled by boring animals. *Size:* Shell length 6-10 in. (150-250 mm). *Habitat:* Attached to intertidal and subtidal rocks; to 60 ft. (18.3 m). *Distribution:* Throughout the Gulf to Peru. *Remarks:* This is one of the largest and heaviest of Panamic bivalves. Large numbers are harvested by commercial and sport divers and their abundance is seriously declining. The lower valve is nearly always cemented to a rock.

64. ***Lima tetrica***

Delicate file shell

Description: The white valves are thin, obliquely oval, and heavily ribbed. *Size:* Shell height 2-3 in. (50-75 mm). *Habitat:* Attached to underside of rocks and in rocky crevices. Intertidal to 100 ft. (30.5 m). *Distribution:* Throughout the Gulf to Ecuador. *Remarks:* This common species attaches itself with byssal threads to rocks. Several individuals often live together within a single crevice.

65. ***Lima pacifica***

Swimming file shell

Description: The shell valves are white, delicate, and somewhat oval in shape. There is a gap at the anterior region of the shells. The sticky, detachable tentacles are readily visible, even when the shell valves are closed. *Size:* Shell height 1-1.5 in. (25-38 mm). *Habitat:* Under intertidal and subtidal rocks; to 150 ft. (46 m). *Distribution:* Throughout the Gulf to southern Mexico. *Remarks:* This species, like *Argopecten circularis,* is capable of swimming away from predators by flapping its valves. It is reported to be unpalatable to some fishes, presumably because the sticky tentacles readily break loose when disturbed and entangle in the predator's mouth.

66. ***Trigoniocardia biangulata***

Gulf cockle

Description: The inflated shell is heavily sculptured with ribs that are serrated where the valves meet. The valves are white with brownish spotting, particularly on small individuals. *Size:* Shell height 2-4 in. (50-100 mm). *Habitat:* In intertidal and subtidal sand and mud; to 150 ft. (46 m). *Distribution:* Throughout the Gulf, at least to Costa Rica. *Remarks:* Large numbers are occasionally trawled by shrimpers.

Class Gastropoda [Snails, Shell-less Snails (Nudibranchs)]

67. ***Diodora inaequalis***

Rough limpet

Description: The shell is dull gray with light to dark brown rays extending from the apical orifice. *Size:* Shell width 1-1.5 in. (25-38 mm). *Habitat:* Attached to the underside of intertidal and subtidal rocks; to 75 ft. (23 m). *Distribution:* Throughout the Gulf to Ecuador.

68. ***Turbo fluctuosus***

Chevron turban shell

Description: The shell color is variable, usually greenish-brown with a brown-and-white chevron pattern along the spiral cords. *Size:* Base diameter 2-3 in. (50-75 mm). *Habitat:* On rocks and crevices in mid-intertidal zones. *Distribution:* Throughout the Gulf to Peru. *Remarks:* Common throughout the Gulf, but becoming uncommon to the south.

69. ***Strombus granulatus***

Knobby fighting conch

Description: The spire and body possess pronounced tubercles. The edge of the outer lip is irregular. The spire is relatively high and slender. *Size:* Shell length 2.5-3 in. (62-75 mm). *Habitat:* In low intertidal and subtidal sand and rubble; most common offshore to 200 ft. (61 m). *Distribution:* Throughout the Gulf to Ecuador. *Remarks:* During the breeding season large numbers congregate inshore and are sometimes exposed at low tide. Juvenile shells vary in color from beige, pink, purple to red, and have a thin, delicate outer lip.

70. ***Strombus gracilior***

Smooth conch

Description: The shell is light tan with a smooth, fairly heavy, reddish lip. The shoulder and spire bear prominent tubercles, but the body whorl is smooth. *Size:* Shell length 2.5-3 in. (62-75 mm). *Habitat:* On intertidal and subtidal sand; to 150 ft. (45 m). *Distribution:* Throughout the Gulf to Peru. *Remarks:* Like *Strombus granulatus,* this common species moves inshore in large numbers to breed. They are often exposed at low tide.

71. ***Strombus galeatus***

Cortez conch

Description: Adult shells are heavy and white with a yellowish-orange aperture. The heavy periostracum is light brown. The lip is thick in adults but thin in some large, yet still immature individuals. *Size:* Shell length 5-10 in. (125-250 mm). *Habitat:* On intertidal and subtidal sand and rubble, and between rocks, occasionally buried; to 150 ft. (45 m). *Distribution:* Throughout the Gulf to Ecuador. *Remarks:* This is one of the largest gastropods in the Gulf and is harvested for food. Known as "burro" in western Mexico, this once abundant conch is rapidly being depleted by commercial fishermen. Large mounds of empty shells often litter beaches near fishing camps.

72. ***Epitonium billeeanum***

Coral wentletrap

Description: The shell is thin, white to light tan with a tan periostracum; the animal is yellow. *Size:* Shell length 0.5-0.75 in. (12-19 mm). *Habitat:* It lives only on stony coral, *Tubastraea coccinea;* shallow subtidal to 250 ft. (76.2 m). *Distribution:* Throughout the Gulf to Ecuador. *Remarks:* The wentletrap feeds primarily on the polyps of *Tubastraea coccinea,* from which it acquires its yellow pigmentation. It is fairly common where *T. coccinea* occurs, usually around offshore islands.

Hans Bertsch

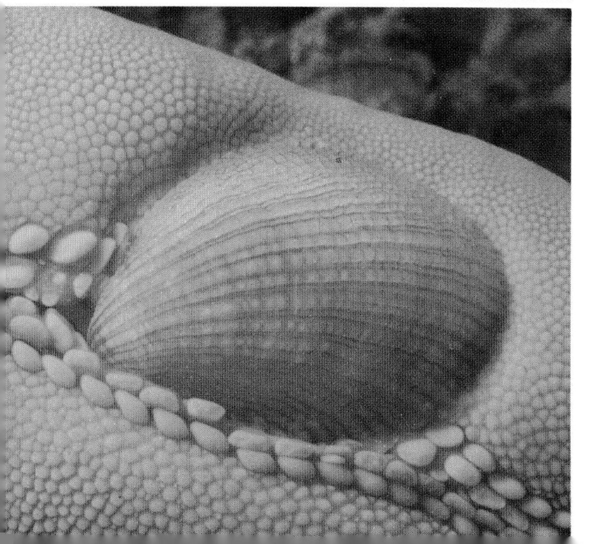

73. ***Thyca callista***

Parasitic cup shell

Description: The shell is tear-drop shaped, with prominent nodulose axial ridges and is creamy-pink. *Size:* Shell length 0.5-0.75 in. (12-19 mm). *Habitat:* On the oral arm surface of the sea stars, *Phataria unifascialis* and *Pharia pyramidata.* Subtidal to depths of 60 ft. (18.3 m). *Distribution:* Central Gulf to Panama. *Remarks:* The large female shell covers the smaller male, and both obtain nourishment by sucking fluids from their sea star host.

74. ***Strombina maculosa***

Stained dove shell

Description: The shell is whitish with brown reticulations. The spire is highly elevated. *Size:* Shell length 1-1.5 in. (25-38 mm). *Habitat:* In or on intertidal and subtidal sand, mud, rubble, and under rocks; to 100 ft. (30.5 m). *Distribution:* Throughout the Gulf to Panama. *Remarks:* This abundant species is often found moving about during the day.

75. ***Polinices bifasciatus***

Striped moonsnail

Description: The whitish bands on the shell distinguish this species from other moonsnails. *Size:* Base diameter 1.5-2 in. (38-50 mm). *Habitat:* In intertidal and subtidal sand; to 150 ft. (46 m). *Distribution:* Throughout the Gulf to Panama. *Remarks:* Like other moonsnails, this common species normally burrows in sand, leaving trails while traveling under it. It is common throughout its range.

76. ***Trivia solandri***

Solander's trivia

Description: The dorsal groove is flanked by whitish nodules that extend as ribs running down to the slit-like aperture. *Size:* Shell length 0.5-0.75 in. (12-19 mm). *Habitat:* Under intertidal and subtidal rocks and in crevices; to 60 ft. (18.3 m). *Distribution:* Throughout the Gulf to Peru, and the outer coast of Baja California from Cabo San Lucas to southern California. *Remarks: T. solandri* is a common species in some localities, particularly along the Sonoran coast. It is often found beached after strong storms or hurricanes.

77. ***Cypraea albuginosa***

Whitespotted cowrie

Description: This species and the Panamic deer cowrie are the only two spotted cowries in the Gulf. The base is lavender when alive but quickly fades to white when dead. The margins are slightly raised. *Size:* Shell length 0.5-1.25 in. (12-32 mm). *Habitat:* Under intertidal and subtidal rocks and in crevices; to 60 ft. (18.3 m). *Distribution:* Central Gulf to Ecuador. *Remarks:* This cowrie is uncommon in the central Gulf but becomes abundant to the south, particularly outside the Gulf.

78. ***Cypraea isabellamexicana***

Orangetip cowrie

Description: The shell dorsum is normally marked with black, often broken streaks. The anterior and posterior ends of the shell are marked with bright orange. *Size:* Shell length 1-2 in. (25-50 mm). *Habitat:* Under intertidal and subtidal rocks and in crevices; to 60 ft. (18.3 m). *Distribution:* Central Gulf to Ecuador. *Remarks:* Although rare in the Gulf, this cowrie is fairly common south of the Gulf. It reaches its greatest size in the Gulf. A similar but smaller species, *Cypraea isabella,* occurs in the Indo-Pacific.

79. ***Cypraea cervinetta***

Panamic deer cowrie

Description: The dorsum is brown to dark mahogany with numerous light gray spots. Some individuals lack spots, particularly juveniles. *Size:* Shell length 3-4 in. (75-100 mm). *Habitat:* Under intertidal and subtidal rocks and in crevices; to 60 ft. (18.3 m). *Distribution:* Throughout the Gulf to Peru. *Remarks:* This species is fairly common in some localities and is the largest cowrie in the Gulf. A dwarf form, 1-1.5 in. (25-38 mm) in length, also occurs in Panama.

80. ***Cypraea annettae***

Annette's cowrie

Description: The shell dorsum is mottled with dark brown to black spots and is usually darker than the base. The teeth are occasionally white and the inside of the shell is purple. *Size:* Length 1.5-2 in. (38-50 mm). *Habitat:* Under intertidal and subtidal rocks and in crevices; to 60 ft. (18.3 m). *Distribution:* Throughout the Gulf to Mazatlán. *Remarks:* This common cowrie lays egg capsules, each containing several hundred eggs. These capsules may vary in color from white, beige, to purple. A subspecies, *Cypraea annettae aequinoctialis,* occurs uncommonly from Panama to Ecuador.

81. ***Cypraea arabicula***

Finetooth cowrie

Description: The shell dorsum is similar to *Cypraea annettae* in color, with an occasional greenish hue. The shell is more tumid, with margins that are sharply angulate. The beige aperture bears numerous fine teeth. *Size:* Shell length 1-1.5 in. (25-38 mm). *Habitat:* Under intertidal and subtidal rocks and in crevices; to 50 ft. (15.2 m). *Distribution:* Central Gulf to Peru. *Remarks:* This species is fairly common throughout its range, and can occasionally be found completely exposed in rock crevices at low tide. Its egg capsules are normally beige to pale yellow.

82. ***Simnia aequalis***

Variable egg shell

Description: Color of the mantle and shell is variable from white, brown to purple. It mimics the polyp color of its gorgonian host. *Size:* Shell length 0.25-0.5 in. (6-12 mm). *Habitat:* Lives commensally on a wide variety of gorgonians including *Eugorgia, Psammogorgia, Muricea,* and others; intertidal and subtidal to 250 ft. (76.2 m). *Distribution:* Throughout the Gulf to Ecuador, and outer Baja California, north from Cabo San Lucas to central California. *Remarks:* A single gorgonian may be host to over 20 individuals of this ovulid cowrie. These snails feed on the polyps of their hosts.

83. ***Jenneria pustulata***

Sea button

Description: The dorsal surface is covered with reddish orange pustules and the ventral surface is banded with raised white ribs. *Size:* Shell length 0.75-1 in. (6-12 mm). *Habitat:* Under intertidal and subtidal rocks and in crevices; to 60 ft. (15.2 m). *Distribution:* Throughout the Gulf to Ecuador. *Remarks:* Common in some localities. Numerous individuals can occasionally be found on beaches after heavy storms or hurricanes. This cowrie-related snail is frequently associated with the colonial anemone, *Palythoa ignota.* Several snails may be found living within a large colony.

84. ***Cassis tenuis***

Thinshell helmet

Description: Shell is quite thin. The aperture is relatively narrow, with a single varix at the outer lip. *Size:* Shell length 3-6.4 in. (75-160 mm). *Habitat:* In subtidal sandy areas close to rocks or under rocky ledges; at depths of 10-150 ft. (3-46 m). *Distribution:* Central Gulf to Ecuador. *Remarks:* A rather rare species which appears to be associated with offshore islands. Crab-inhabited shells are occasionally found, but live specimens are rarely seen.

85. ***Cassis coarctata***

Narrow helmet shell

Description: Smaller and heavier than *Cassis tenuis*. The upper part of the lip folds slightly inward. *Size:* Shell length 1.5-3 in. (37-75 mm). *Habitat:* In intertidal and subtidal sand, between rocks; to depths of 100 ft. (30.5 m). *Distribution:* Throughout the Gulf to Ecuador. *Remarks:* Remains buried under the sand during the day, but emerges at night to forage over sand. It is locally common.

86. ***Cassis centiquadrata***

Spotted bonnet shell

Description: Ovate shell with brownish spots. The aperture is wide and brown. *Size:* Shell length 2-3 in. (50-75 mm). *Habitat:* Buried in intertidal and subtidal sand; to depths of 200 ft. (61 m). *Distribution:* Throughout the Gulf to Peru. *Remarks:* Appears to be more common offshore; often caught in large numbers by shrimp trawlers.

87. ***Morum tuberculosum***

Yellowmouth helmet

Description: Grayish brown shell with a pale yellowish aperture. *Size:* Shell length 0.75-1.5 in. (19-38 mm). *Habitat:* Under rocks in intertidal and subtidal sand habitats; to 60 ft. (18.3 m). *Distribution:* Throughout the Gulf to Peru. *Remarks:* Locally common.

88. ***Ficus ventricosa***

Panamic fig shell

Description: The shell is very thin, usually tan with a very wide aperture. *Size:* Shell length 3-5 in. (75-125 mm). *Habitat:* Offshore sandy habitat; to depths of 400 ft. (122 m). *Distribution:* Throughout the Gulf to Peru. *Remarks:* Often caught by shrimp trawlers in large numbers. The animal burrows in the sand backward using the posterior end of the foot to dig into the sand.

89. ***Distorsio constricta***

Bearded distorsio

Description: The anterior canal is "bent" to the right of the line of symmetry. A fold is present on the upper part of the inner lip. The periostracum is covered with heavy tufts of hair. *Size:* Shell length 1.5-3 in. (38-75 mm). *Habitat:* In sand; mostly offshore at depths of 55-250 ft. (16.8-76.2 m). *Distribution:* Throughout the Gulf to Peru. *Remarks:* This common shell is often captured by shrimp trawlers.

90. ***Bursa sonorana***

Sonoran frog shell

Description: Spire has sharp nodes; the shell is light tan to white with brown bands. *Size:* Shell length 3-4 in. (75-100 mm). *Habitat:* In sand; offshore at depths of 100-300 ft. (30.5-91.4 m). *Distribution:* Throughout the Gulf. *Remarks:* This common frog shell is frequently caught by shrimp trawlers.

91. ***Murex tricoronis* (formerly *Murex recurvirostris tricoronis*)**

Longspine murex

Description: The shell has sharp spines and is light brown, tan to white, with a number of fine reddish-brown bands. *Size:* Shell length 2-3.5 in. (50-87 mm). *Habitat:* Offshore, in sand, at depths of 55-400 ft. (16.8-122 m). *Distribution:* Throughout the Gulf to Ecuador. *Remarks:* This common species is most often found as a byproduct of shrimp trawl material. Occasionally, large sponge-like egg masses may contain 10-30 adult animals buried within. Three subspecies are currently recognized.

92. ***Chicoreus erythrostomus* (formerly *Hexaplex erythrostomus*)**

Pink murex

Description: The aperture is brilliant pink to deep scarlet. Deep water individuals have longer spines. *Size:* Shell length 4-6 in. (100-150 mm). *Habitat:* In intertidal and subtidal sand and on rubble; to depths of 250 ft. (76.2 m). *Distribution:* Throughout the Gulf to at least Costa Rica. *Remarks:* This species is the most common murex in the Gulf. Occasionally some specimens may have completely white apertures, or bear only a faint pink border on the edge of the outer lip. During breeding season they may congregate in large numbers, sometimes a hundred or more. They feed heavily on the winged pearl oyster, *Pteria sterna*.

93. ***Pterynotus pinniger***

Webbed murex

Description: The spines are modified into delicate webbed fronds. The shell color is variable and often bicolored, white, pink, lavender, or brown. *Size:* Shell length 1.3-3 in. (38-75 mm). *Habitat:* In rubble, occasionally burrowing in sand. Offshore and off islands; subtidal at depths of 60-250 ft. (18.3-76.2 m). *Distribution:* Central Gulf to Ecuador. *Remarks:* Adult specimens of this rare species are usually heavily encrusted and often drilled by worms.

94. ***Pteropurpura centrifuga***

Centrifugal murex

Description: The shell is yellowish-tan overall. The tip of the spines curve slightly in a counterclockwise direction. Moderate webbing between spines is evident in most specimens. *Size:* Shell length 2-3.5 in. (50-88 mm). *Habitat:* Offshore in sand; at depths of 100-500 ft. (30.5-152.4 m). *Distribution:* Throughout the Gulf to Panama. *Remarks:* This uncommon species is occasionally caught by shrimp trawlers.

95. ***Pleuroploca princeps*** **(formerly *Fasciolaria princeps*)**

Panamic horse conch

Description: The shell is light tan and covered with a heavy brown periostracum. The animal is brilliant red, dotted with blue spots. *Size:* Shell length 6-18 in. (150-450 mm). *Habitat:* Among rocks at extreme low tide; subtidal to 200 ft. (61 m). *Distribution:* Throughout the Gulf to Peru. *Remarks:* This relatively common species is the largest of Panamic gastropods. It preys on a variety of large gastropods such as *Murex* and *Strombus*. Like *Strombus galeatus,* this conch is also harvested for food by commercial divers.

96. ***Harpa crenata***

Crenate harp shell

Description: The shell is semi-glossy with prominent axial ribs. The aperture is wide and the columellar margin is glossy. *Size:* Shell length 3-3.5 in. (75-87 mm). *Habitat:* In subtidal sand habitats, usually offshore at depths of 75-400 ft. (22.9-122 m). *Distribution:* Throughout the Gulf to Ecuador. *Remarks:* This species, like other harp shells, can autotomize the posterior portion of its foot. Although rare inshore, it is commonly trawled by shrimpers offshore.

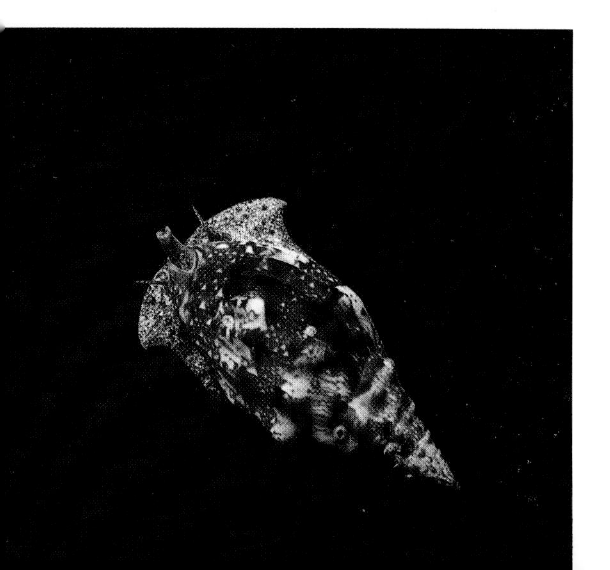

97. ***Lyria cumingi***

Cuming's lyria

Description: The spire is elevated with nodose ribbing. *Size:* Shell length 1.25-1.5 in. (32-38 mm). *Habitat:* In intertidal and subtidal sand or mud habitats and under rocks; to 100 ft. (30.5 m). *Distribution:* Throughout the Gulf to Peru. *Remarks:* This abundant species is one of only two members of the volute family to occur in the Panamic province. The other, *Lyria barnessi,* is uncommon, occurs offshore and is distributed from the central Gulf to Peru. It is more ovate, with less nodose ribbing than *Lyria cumingi.*

98. ***Falsifusus dupetitthouarsi***
(formerly *Fusinus dupetitthouarsi*)

Giant spindle shell

Description: The spire and canal are nearly equal in length. The periostracum is thick, almost "woolly" and yellowish-brown. *Size:* Shell length 5-10 in. (125-250 mm). *Habitat:* In sand and rubble at extreme low tide and subtidally to 300 ft. (91.4 m). *Distribution:* Throughout the Gulf to Ecuador. *Remarks:* Although rare inshore, large numbers of individuals are nearly always caught in trawls by shrimpers.

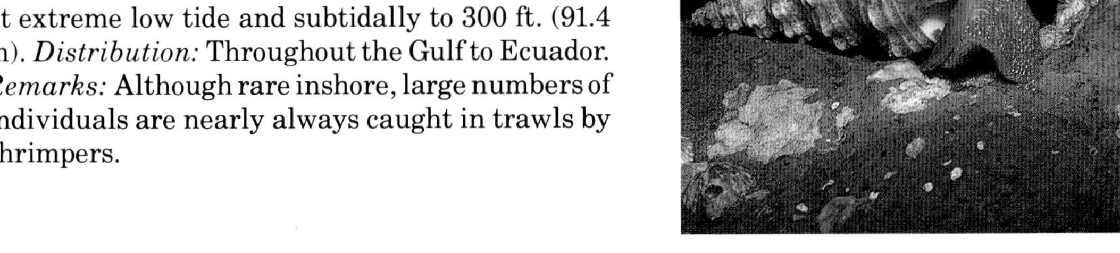

99. ***Oliva incrassata***

Angulate olive shell

Description: Mature specimens have an unusually thick outer lip. The shoulder area is particularly bulbous. Shells are gray, tan, or brown with broken vertical irregular bands. The aperture may be yellowish-pink. *Size:* Shell length 2-3.5 in. (50-87 mm). *Habitat:* In intertidal and subtidal sandy regions; to 250 ft. (76.2 m). *Distribution:* Throughout the Gulf to Peru. *Remarks:* The color of this common shell is highly variable. There are two other color forms found only in the San Felipe area of Baja California, a rare white form and an uncommon golden form. Although a black form has been reported, none has been documented thus far.

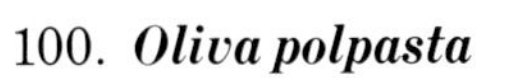

100. ***Oliva polpasta***

Bulbous olive shell

Description: The shell is gray to brown with fine brown spotting, similar to *Oliva spicata* but more ovate. *Size:* Shell length 1.25-1.5 in. (32-38 mm). *Habitat:* In low intertidal and subtidal sand; to 200 ft. (61 m). *Distribution:* Throughout the Gulf to Ecuador. *Remarks:* Although the color of this common species is variable, it is less so than *Oliva incrassata* and *O. spicata*. The rare banded olive, *Oliva kerstitchi*, may only be a color form of *Oliva polpasta*.

101. ***Oliva porphyria***

Tent olive

Description: The glossy shell is covered with a network of striking brown tent markings and can be confused with no other species of olive shell. *Size:* Shell length 3-5 in. (75-125 mm). *Habitat:* In sand at extreme low tide and subtidally to 200 ft. (61 m). *Distribution:* Throughout the Gulf to Panama. *Remarks:* This relatively common species is the world's largest olive. Early voyagers occasionally mistook this shell for the famed Glory-of-the-Sea cone, *Conus gloriamaris.* It feeds on small gastropods, including other olives.

102. ***Oliva spicata***

Veined olive shell

Description: The shell color varies from light brown to gray or almost white. The pattern consists of small irregular tent markings or spots, entirely covering the shell on some specimens. The spire is usually high. *Size:* Shell length 1-3 in. (25-75 mm). *Habitat:* In intertidal and subtidal sand; to 200 ft. (61 m). *Distribution:* Throughout the Gulf to Panama. *Remarks:* Like all other olives, this species leaves behind undulating trails in the sand when foraging about for food at night. A number of forms and subspecies are recognized.

103. ***Agaronia testacea***

Panamic slender olive shell

Description: The shell color is variable from gray to bluish-gray or buff, with brown axial lines. Unlike the closely related genus *Oliva, Agaronia* has an operculum. *Size:* Shell length 1.25-2.5 in. (32-62 mm). *Habitat:* Often in estuaries, buried in sand; intertidal and subtidal to 100 ft. (30.5 m). *Distribution:* Throughout the Gulf to Peru. *Remarks:* As do other olives, this species leaves tracks in the sand when foraging at night. During low tide it may be completely exposed for several hours.

104. ***Olivella dama***

Pigmy olive

Description: The whitish shell has a broad gray to brown band, and irregular tent markings. The aperture is violet. *Size:* Shell length 0.5-0.75 in. (12-19 mm). *Habitat:* In intertidal and subtidal sand; to below 200 ft. (61 m). *Distribution:* Throughout the Gulf to Mazatlán. *Remarks:* At low tide this abundant species is often stranded out of water, leaving distinct tracks as it moves about in wet sand.

105. ***Conus brunneus***

Brown cone shell

Description: The shell is solid brown with cream-white patches. The spire has incised grooves. *Size:* Shell length 2-3 in. (50-75 mm). *Habitat:* Under intertidal and subtidal rocks or crevices; to 80 ft. (24.4 m). *Distribution:* Throughout the Gulf to Ecuador. *Remarks:* Juveniles under 0.75 in. (19 mm) are white with dark brown axial bands. This common species is vermivorous, feeding only on polychaete worms.

106. ***Conus princeps***

Princely cone shell

Description: The shell is pale orange to reddish orange. The black axial bands are sometimes broken. The profile of the lip is slightly scalloped. *Size:* Shell length 2-3.5 in. (50-87 mm). *Habitat:* Under intertidal and subtidal rocks or exposed on rubble; to 100 ft. (30.5 m). *Distribution:* Throughout the Gulf (endemic to the Gulf). *Remarks:* *Conus princeps* is relatively common in the Gulf. Two color variants occur in Panama: *Conus princeps* var. *lineolatus,* with hair-like axial bands, and *Conus princeps* var. *apogrammatus,* which is solid orange, without any axial bands. This species is vermivorous, feeding only on polychaete worms.

107. *Conus purpurascens*

Purple cone

Description: The shell is light to dark purple, with an irregular pattern of brown or black streaks and blotches. *Size:* Shell length 1.5-3 in. (38-75 mm). *Habitat:* Under intertidal and subtidal rocks, buried in sand; to 50 ft. (15.2 m). *Distribution:* Throughout the Gulf to Ecuador. *Remarks:* This common piscivorous species feeds primarily on small reef fishes. Like other piscivorous species, this cone should be handled carefully, as its sting is venomous.

108. *Conus dalli*

Dall's cone

Description: The shell has a network of tent-like and triangular brown and yellow markings. The aperture is pale rosy pink. *Size:* Shell length 2-3 in. (50-75 mm). *Habitat:* Buried in intertidal and subtidal sand or under rocks; to 90 ft. (27.4 m). *Distribution:* Central Gulf to Ecuador. *Remarks:* This uncommon molluscivorous species feeds mostly on small snails, including other cones. It is venomous and potentially dangerous if handled with bare hands since it may sting when disturbed.

109. *Conus scalaris*

Stepspire cone shell

Description: A high spired, white, slender shell covered with rusty to dark mahogany flammules. *Size:* Shell length 1.5-2.5 in. (38-64 mm). *Habitat:* Offshore in sand; at depths of 50-400 ft. (15.2-122 m). *Distribution:* Throughout the Gulf to Ecuador. *Remarks:* This uncommon species is highly variable, both in shell morphology and coloration.

110. ***Conus fergusoni***

Giant cone shell

Description: Adult shells are normally pure white. Juvenile shells under 2 in. (50 mm) have yellow to bright orange bands on the body whorl. Occasionally adult shells will retain faint orange bands. A heavy, brown periostracum covers the shell. *Size:* Shell length 3-7 in. (75-175 mm). *Habitat:* On subtidal sand, patch reefs and rubble; at depths of 15-500 ft. (4.6-152.4 m). *Distribution:* Throughout the Gulf to Ecuador. *Remarks:* This is the largest of the eastern Pacific cones. Although often captured offshore in shrimp trawls, it is uncommon inshore.

111. ***Conus xanthicus***

Midas cone shell

Description: The shell is yellow to orange, broken by a light, midbody spiral band, becoming irregular in mature specimens. The periostracum is fairly thick in adults, bearing "tufts" along the shoulder. *Size:* Shell length 1.5-2.5 in. (38-63 mm). *Habitat:* Offshore, in sand and rubble; at depths of 100-400 ft. (30.5-122 m). *Distribution:* Central Gulf to Ecuador. *Remarks:* The adult of this rare species was, until recently, confused with young *Conus fergusoni*. The shoulder of *Conus xanthicus* is angular, rounder in *C. fergusoni*. The spire of C. *xanthicus* is marked with orange flammules that are never present in *C. fergusoni*.

112. ***Conus nux***

Pigmy cone shell

Description: The brown markings form two or more indistinct bands. The live animal is always pink. *Size:* Shell length 0.5-1 in. (12-25 mm). *Habitat:* Exposed on rocky intertidal and subtidal substrates; to 40 ft. (12.2 m). *Distribution:* Throughout the Gulf to Ecuador. *Remarks:* This is the smallest and most common of the eastern Pacific cones. It is one of the few species which forages for food during the day.

113. ***Conus ximenes***

Finespot cone shell

Description: The profile of each spire whorl is moderately concave, as opposed to the straighter or convex spire whorl of the similar *Conus mahogani.* Fresh specimens possess a bluish to lilac tinge over the entire shell. *Size:* Shell length 1.5-2 in. (38-50 mm). *Habitat:* In sand, near the mouth of estuaries and offshore; to 300 ft. (91.4 m). *Distribution:* Throughout the Gulf to Peru. *Remarks:* This species is relatively common locally. It is vermivorous, feeding on small annelid worms. A pustulose form occurs in Panama.

114. ***Conus bartschi***

Bartsch's cone shell

Description: The shell is dull white with irregular brown blotches. Fine brown dots circle the entire body whorl. It lacks the spire grooves found in *C. brunneus.* The live animal is brilliant red (purple in *C. brunneus). Size:* Shell length 1.25-2 in. (32-50 mm). *Habitat:* Exposed on subtidal rubble and among rocks; at depths of 20-250 ft. (6.1-76.2 m). *Distribution:* Central Gulf to Panama. *Remarks:* Once thought to be a deep water color variation of *Conus brunneus,* this rare species has proven to be distinct. Exposed individuals are nearly always heavily encrusted. This species is vermivorous, feeding on polychaete worms.

115. ***Terebra strigata***

Zebra auger shell

Description: The shell is usually white, marked with brownish mahogany axial streaks that are interrupted or broken at each whorl. *Size:* Shell length 3.5-5 in. (87-125 mm). *Habitat:* Buried in intertidal and subtidal sand; to 200 ft. (61 m). *Distribution:* Throughout the Gulf to Ecuador. *Remarks:* This auger is common throughout its range. Occasionally some specimens are entirely chestnut-brown. Pure white specimens lacking the dark brown axial stripes have been reported, but are rare.

116. ***Navanax inermis***

Navanax

Description: The large parapodial folds extend along the sides of the animal, nearly meeting mid-dorsally. The body is tan to purple or black with yellowish streaks; a yellow or orange band and a row of electric blue dots line the margins. A small, calcified shell is buried beneath the dorsal mantle surface. There is no radula. *Size:* Body length 2.5-10 in. (63-250 mm). *Habitat:* On mudflats and sandy bays; subtidally. *Distribution:* Throughout the Gulf to Panama; the outer coast of Baja California to central California. *Remarks:* This common sea slug is a voracious predator of other opisthobranchs (especially *Bulla* and *Haminoea*) actively tracking its prey by following slime trails left by prey on the sand. It swallows its prey whole. Often it can be found "resting" inside a sand-coated mucus tube.

117. ***Tridachiella diomedea***

Mexican dancer

Description: The parapodial folds are frilled. The body is greenish with longitudinal black and yellow lines on rhinophores and light blue and yellow lines on the edge of the body folds. *Size:* Body length 1.25-2 in. (32-50 mm). *Habitat:* On intertidal and subtidal rocks and reefs; to 75 ft. (22.9 m). *Distribution:* Throughout the Gulf to Panama. *Remarks:* One of the most common opisthobranchs in the Gulf of California. The greenish body coloration results from the presence of symbiotic chloroplasts in its tissues.

118. ***Aplysia californica***

California sea hare

Description: The large, humped body has dorsal-projecting flaps on each side. A spacious mantle cavity is formed by these thick, flexible flaps. The partly calcified shell is flattened and earlike in shape. *Size:* Body length 6-30 in. (150-750 mm). *Habitat:* On intertidal and subtidal sand and rocks; to 60 ft. (18.3 m). *Distribution:* Throughout the Gulf; along the outer coast of Baja California to northern California. *Remarks:* These common sea hares emit a purple defensive secretion when disturbed. They deposit long, tangled skeins of yellow, yellow-green, or pink eggs. Each individual is cap-

able of laying over 750,000 eggs in one season. They eat various species of red and green algae.

119. ***Dolabella auricularia***

Bluntend sea hare

Description: The posterior end of the triangular body is typically truncate. A well-developed, calcified internal shell is present. The body color varies from dark brownish green to nearly black. *Size:* Body length 4-8 in. (100-200 mm). *Habitat:* Among intertidal and subtidal rocks, near algae; to 50 ft. (15.2 m). *Distribution:* Central Gulf to Ecuador; circumtropical. *Remarks:* This locally common animal produces physiologically active chemical compounds that are being evaluated for their use in cancer chemotherapy.

120. ***Tylodina fungina***

Yellow umbrella shell

Description: The brilliant yellow body has its gill along the right side. The external shell is thin, transparent, and marked with brown-black rectangles peripherally. *Size:* Body length 0.75-1.5 in. (19-38 mm). *Habitat:* Under intertidal and subtidal rocks; to at least 60 ft. (18.3 m). *Distribution:* Throughout the Gulf to Ecuador, and north along Baja to southern California. *Remarks:* Usually occurs on its similarly-colored prey sponge, *Aplysina fistularis,* where it is well camouflaged. It is seasonally common.

121. ***Berthellina engeli***

Apricot slug

Description: The body color is variable from pale yellow to orange or red. The small, flat shell is internal. The dorsum is smooth with the gill located on the right side of the body. *Size:* Body length 1.5-2.5 in. (19-38 mm). *Habitat:* Under intertidal and subtidal rocks; to 100 ft. (30.5 m). *Distribution:* Throughout the Gulf to Ecuador, and southern California. Also reported from the temperate coasts of the east Atlantic. *Remarks:* Individuals of this species often occur commonly in pairs under rocks. The notal skin can secrete defensive sulfuric acid.

122. ***Pleurobranchus areolatus***

Warty sea slug

Description: The dorsal surface is covered with numerous large and small, bubble-like tubercles. The body is brown orange to brick red, with white patches. *Size:* Body length 2-6 in. (50-150 mm). *Habitat:* Under intertidal and subtidal rocks; to 80 ft. (30.5 m). *Distribution:* Throughout the Gulf to Ecuador, also throughout the Caribbean and tropical west Africa. *Remarks:* A relict species living on both coasts of Central America, it has not speciated since the closure of the Tertiary Panamic-Central American seaway.

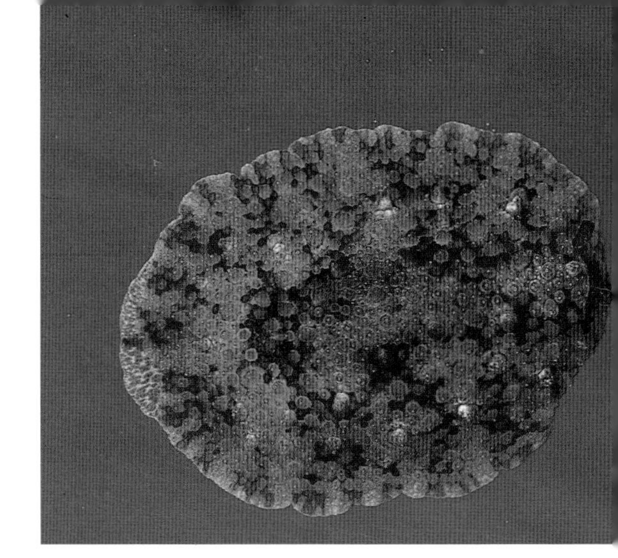

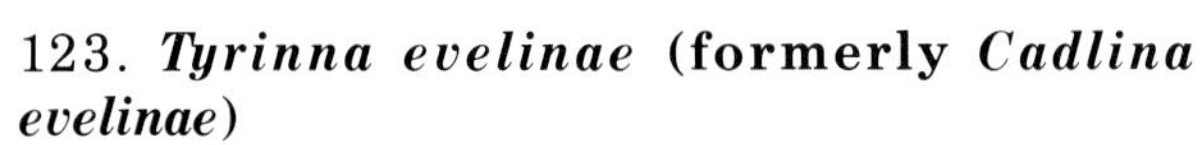

123. ***Tyrinna evelinae* (formerly *Cadlina evelinae*)**

Eveline's sea slug

Description: The soft, white body has numerous small, bright red to yellow orange spots. *Size:* Body length 0.75-1.25 in. (19-32 mm). *Habitat:* Under intertidal and subtidal rocks. *Distribution:* Throughout the Gulf; also reported from Brazil. *Remarks:* Another Panamic species of nudibranch that has a tropical amphi-American distribution.

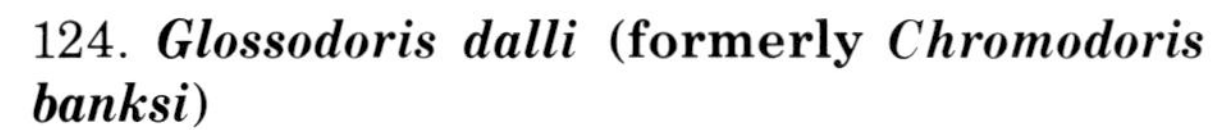

124. ***Glossodoris dalli* (formerly *Chromodoris banksi*)**

Dall's nudibranch

Description: White to cream background with brown-black spots of varying size on the dorsal surface, and sides of the body. Larger specimens tend to have more numerous, smaller black dots and the background becomes grayish around the center of the dorsum. Cream-colored, and occasionally orange spots (red in larger specimens) are mixed among the blackish dots on the dorsum. The gills and rhinophores are white, tipped with orange or light red. *Size:* Body length 0.75-1.75 in. (19-44 mm). *Habitat:* Under intertidal and subtidal rocks. *Distribution:* Throughout the Gulf, to at least Costa Rica. *Remarks:* This common nudibranch feeds on sponges, which it grasps with its radula.

125. ***Glossodoris sedna* (formerly *Casella sedna*)**

Sedna

Description: The margin of the white body is encircled by three colored bands of off-white, red, and yellow. The rhinophores and gill tips are red. *Size:* Body length 1.5-2.5 in. (38-63 mm). *Habitat:* Under intertidal and subtidal rocks; to at least 60 ft. (18.3 m). *Distribution:* Throughout the Gulf to Ecuador. *Remarks:* The gills nearly always vibrate and wiggle back and forth. The chemicals found in the defensive secretions of dorids are generally identical with or similar to the metabolites of their prey sponges. Researchers have isolated several tetracyclic terpene compounds from this species, including the compound sednolide.

126. ***Chromodoris annulata***

Annulated nudibranch

Description: White body with yellow spots, a dark purple margin, and a dark purple ring around the gills and both rhinophores. *Size:* Body length 1-1.25 in. (38-63 mm). *Habitat:* Under intertidal and subtidal rocks. *Distribution:* Reported from two disjunct localities; before the single record from the Gulf (Isla Tortuga), this species had been reported from the eastern coast of Africa and eastern Australia. *Remarks:* Additional records from the Gulf would be noteworthy.

127. ***Chromodoris baumanni***

Baumann's nudibranch

Description: Numerous small, red-purple dots cover the dorsum and the posterior and lateral surfaces of the foot. An interrupted band of orange dots and streaks surrounds the body margin and foot. The rhinophores are white, red-purple distally. The gills are white, with a purplish hue on the distal portion. *Size:* Body length 2-2.5 in. (50-63 mm). *Habitat:* Under intertidal and subtidal rocks; to 60 ft. (18.3 m). *Distribution:* Central Gulf to Ecuador. *Remarks:* A similarly-colored unnamed species can be distinguished by its green markings under the notal margin.

128. ***Chromodoris galexorum***

Galactic sea slug

Description: The brilliant scarlet splotches are outlined with a chrome-yellow ring; gills and rhinophores are blood red. The body margin is rimmed dorsally with a thin, yellow-chrome line. *Size:* Body length 0.75-1.5 in. (19-38 mm). *Habitat:* Under subtidal rocks and crevices; at depths of 10-150 ft. (3-45.7 m). *Distribution:* Central and southern Gulf (including one record from Isla Guadalupe off the outer coast of Baja). *Remarks:* The Isla Guadalupe record may be a thermally anomalous occurrence, outside the animal's normal distribution because of elevated water temperatures during a prolonged El Niño period.

129. ***Chromodoris marislae***

Marisla's dorid

Description: Body color is usually off-white, with two or three irregular rows of orange spots encircling the periphery of the body; central to these rows is a roughly circular arrangement of larger orange ringlets, which are often surrounded or marked centrally with pure white. *Size:* Body length 2.5-3 in. (63-75 mm). *Habitat:* Under rocks and crevices; subtidally at depths of 15-100 ft. (4.6-30.5 m). *Distribution:* Central and southern Gulf (distribution outside the Gulf is unrecorded). *Remarks:* The egg mass of this uncommon species is white, gently ruffled on the free edge, with about 2½ whorls. The chemical defense of *Chromodoris marislae* is primarily a metabolite, known as marislin.

130. ***Chromodoris norrisi***

Clown nudibranch

Description: There is a wide range of variation in the number and size of the red and yellow spots, from many small ones, to a few large ones. *Size:* Body length 1.5-2.5 in. (38-63 mm). *Habitat:* Under intertidal and subtidal rocks and crevices; to 50 ft. (15.2 m). *Distribution:* Throughout the Gulf and up the outer Baja coast to Isla Cedros. *Remarks:* Locally and seasonally abundant. This species contains an unusual diterpene compound (called norrisolide) that may function as a chemical defense mechanism. The chemical is probably a biosynthetic product from a compound derived from its sponge prey.

131. *Chromodoris sphoni*

Red cross nudibranch

Description: The dorsal red-cross pattern is characteristic of this species. *Size:* Body length 0.5-1.25 in. (12-32 mm). *Habitat:* Under and upon intertidal and subtidal rocks; to 60 ft. (18.3 m). *Distribution:* Central Gulf to Panama. *Remarks:* Animals collected in Panama fed on a pink-red sponge.

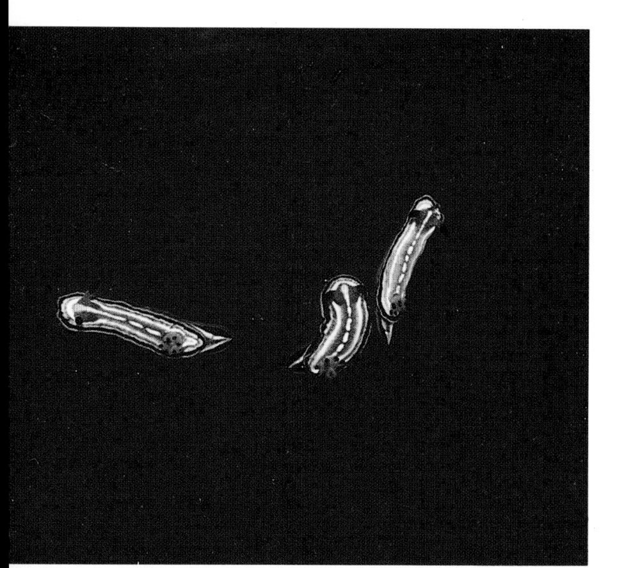

132. *Mexichromis antonii*

Antonio's nudibranch

Description: Body color consists of shades of blue, magenta, black, yellow-orange, and white. A yellow-orange line encircles the body margin; a black line immediately borders the inner side of the yellow-orange band. A wide area of light blue covers the rest of the lateral body region. *Size:* Body length 0.25-0.50 in. (6-12 mm). *Habitat:* Under and upon rocks; subtidally to 80 ft. (24.4 m). *Distribution:* Central Gulf, at least to Costa Rica. *Remarks:* This relatively rare chromodorid has been collected from several scattered locations throughout the Panamic Province.

133. *Hypselodoris agassizii*

Agassiz nudibranch

Description: Body color is blue-black, sprinkled with numerous small yellow dots. At times there are also elongated, ovalish white marks on the dorsum. The colored stripes run along the body margin: a yellow band on the outside, with a narrower navy blue or black band in the middle, and a light green one innermost. *Size:* Body length 1.25-1.50 in. (32-38 mm). *Habitat:* On and under intertidal and subtidal rocks; to at least 60 ft. (18.3 m). *Distribution:* Throughout the Gulf to Ecuador. *Remarks:* The color pattern of interrupted stripes at the middle of the body may serve as disruptive coloration in the animal's normal habitat.

134. ***Hypselodoris californiensis***

California dorid

Description: The yellow to yellow-orange, round, or oblong spots usually run in several longitudinal rows on the dorsum with another row along each side of the foot. *Size:* Body length 1.5-2.5 in. (38-63 mm). *Habitat:* Under and upon intertidal and subtidal rocks; to 90 ft. (27.4 m). *Distribution:* Central Gulf (distribution in the Gulf is not well documented); outer Baja California from Bahía Magdalena to Monterey Bay, California. *Remarks:* Radular and external morphology distinguish *H. californiensis* from *H. ghiselini* (see comparative remarks under that species).

Hans Bertsch

135. ***Hypselodoris ghiselini***

Ghiselin's sea slug

Description: Body color is deep navy blue, covered with numerous bright yellow specks. Rhinophores and gills are navy blue with yellow dots on the inner sides of the gills. Very similar to *Hypselodoris californiensis,* but the blue-black color is much darker in *H. ghiselini,* with more numerous, smaller yellow dots. *Hypselodoris californiensis* has larger yellow maculations, and a light bluish-white tinge around the edge of the notum. *Size:* Body length 2-2.75 in. (50-70 mm). *Habitat:* On and under intertidal and subtidal rocks. *Distribution:* Throughout the Gulf (may be a Gulf endemic). *Remarks:* This uncommon sea slug secretes metabolites (obnoxious chemicals for defense against predators).

Hans Bertsch

136. ***Sclerodoris tanya***

Tanya's dorid

Description: The dorsum has large, deep pits and irregular tubercles. The body color is tan with dark tan or brown spots. *Size:* Body length 1.75-2 in. (44-50 mm). *Habitat:* On rocky intertidal substrates. *Distribution:* Throughout the Gulf, and the outer coast of Baja California to southern California. *Remarks:* This is the only species of *Sclerodoris* known from the eastern Pacific; the other species of *Sclerodoris* are Indo-Pacific (from Africa to Hawaii and Japan).

Hans Bertsch

Hans Bertsch

137. ***Taringa aivica timia***

Gulf Taringa

Description: Body color is dusky yellow with light frosting on the dorsum. Elongated papillae occur on the dorsum. *Size:* Body length 2.5-2.75 in. (63-70 mm). *Habitat:* Not well known. *Distribution:* Upper Gulf (distribtuion throughout the Gulf is undetermined), and southern California. *Remarks:* Further anatomical studies may give the data necessary to raise this subspecies to full specific status.

Hans Bertsch

138. ***Aegires albopunctatus***

White knight

Description: The body is rigid, with a spiculose skin texture. Irregular rows of tubercles are present on the notum. *Size:* Body length 0.75-1 in. (19-25 mm). *Habitat:* Rocky intertidal and subtidal. *Distribution:* Central Gulf (reported only from Bahía de los Angeles) and outer coast of Baja California to British Columbia. *Remarks:* This species feeds on sponges of the class Calcarea (e.g., *Leucilla* and *Leucetta*).

139. ***Tambja eliora* (formerly *Nembrotha eliora*)**

Blue striped sea slug

Description: Body color is green to yellow ochre; three turquoise blue stripes on the dorsum are bordered by black. Rhinophores and gills are deep blue-black, the gills with bluish or greenish central stalks. *Size:* Body length 1-2 in. (25-50 mm). *Habitat:* Over intertidal and subtidal rocky substrates; to 150 ft. (45.7 m). *Distribution:* Throughout the Gulf (distribution outside the Gulf is not well known). *Remarks:* This common species can swim away from predators by undulating its body from side to side to propel itself through the water.

140. ***Tambja abdere***

Slimy slug

Description: The body is turquoise with black-rimmed patches of light green and is rimmed with a light green band edged with a thin black line. The rhinophores are blue-black. The dark gills are set on yellow ochre stalks. *Size:* Body length 2.5-3.25 in. (63-82 mm). *Habitat:* On subtidal rocky substrates; at depths of 10-200 ft. (3-61 m). *Distribution:* Central and southern Gulf, outer Baja California from Cabo San Lucas to Bahía Magdalena. *Remarks:* This uncommon sea slug secretes copious amounts of mucus when attacked by its predator, the sea tiger nudibranch, *Roboastra tigris*.

141. ***Roboastra tigris***

Sea tiger

Description: The basic body color is green to yellow-ochre, with five longitudinal black stripes outlined by light green. The rhinophores are blue-black and the gills are blue-black with light green central axes. The oral area is cobalt blue, with two lateral inrolled dark blue processes (probably chemosensory or tactile organs). *Size:* Body length 3-12 in. (75-300 mm). *Habitat:* On rocks, at depths of 25-200 ft. (7.6-61 m). *Distribution:* Central and southern Gulf (may be a Gulf endemic). *Remarks:* A truly impressive predator. Carnivorous on other nudibranchs (including its own species), the sea tiger swallows its prey whole, using large hook-like radular teeth to pull the animal inside its mouth. This species is locally common.

142. ***Polycera alabe***

Inkstain nudibranch

Description: The body is elongated with non-retractile gills. The papillar velar and notal processes are transclucent or black. *Size:* Body length 0.75-1 in. (19-25 mm). *Habitat:* On bryozoans in intertidal and subtidal habitats. *Distribution:* Throughout the Gulf and Isla Cedros off outer Baja California. *Remarks:* There are a number of opisthobranchs in the Gulf that have dark blue (or brownish) bodies covered with yellow or orange spots. Most species tend not to co-occur, due to rather different ecological preferences.

143. ***Trapania* n. sp.**

Yellowtip trapania

Description: White body with distinct chocolate-brown streaks and patches. The tips of the cephalic tentacles, rhinophores, gills, lateral extrabranchial appendages, and the posterior extreme of the foot are a brilliant yellow. *Size:* Body length 0.75 in. (19-25 mm). *Habitat:* On intertidal and subtidal rocky substrates with bryozoans. *Distribution:* Central and southern Gulf, outer coast of Baja California at Isla Cedros and Bahía Sebastian Vizcaino. *Remarks:* This rare species is currently being described.

144. ***Dendrodoris krebsii***

Black dorid

Description: The velvety black animal has a red line around the ruffled body margin. The tips of the gills and rhinophores are white. *Size:* Body length 2-2.5 in. (50-63 mm). *Habitat:* Under intertidal rocks; to below 50 ft. (15.2 m). *Distribution:* Throughout the Gulf to southern Mexico, also west Atlantic, from Florida to Brazil. *Remarks:* Apparently this common species existed before the late Pliocene-early Pleistocene closing of the Central American seaway.

145. ***Doriopsilla albopunctata***

White spotted sea goddess

Description: The dorsum is brownish orange, fading to yellow-orange marginally; small white dots are densely packed over the surface. Rhinophores and gills are yellowish-light orange. *Size:* Body length 2-2.5 in. (50-63 mm). *Habitat:* Intertidal and subtidal rocky habitats. *Distribution:* Throughout the Gulf and outer Baja California from Punta Eugenia to Mendocino, California. *Remarks:* Differences between the similar species, *Dendrodoris fulva* and *D. albopunctata,* are subtle and based mainly on internal anatomy. Positive identification of these species may require dissection.

146. ***Doriopsilla janaina***

Knobby sea slug

Description: The body is light red to bright pink, with two ill-defined blackish bands along the mid-dorsal region. Dorsal tubercles have varying amounts of red and white color. The gills are red. *Size:* Body length 1 in. (25 mm). *Habitat:* Under intertidal rocks. *Distribution:* Throughout the Gulf to Panama. *Remarks:* Although it has a large range, this species has been recorded from only a few collecting localities.

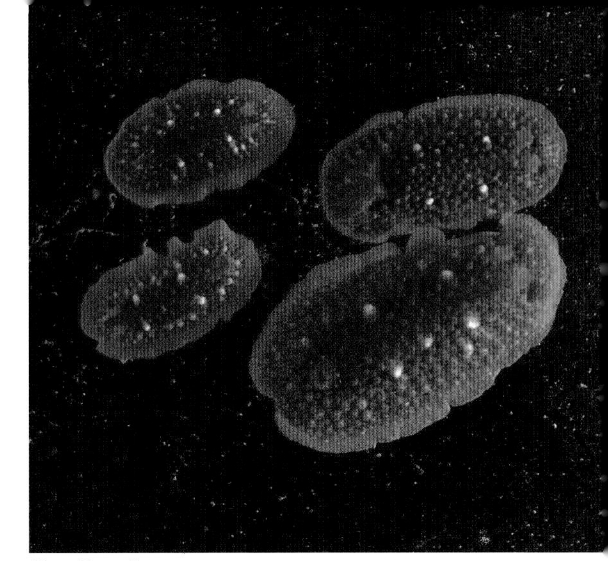

Hans Bertsch

147. ***Tritonia diomedea***

Diomedes' triton

Description: Branchial appendages are located around the edge of the mantle, and a frontal veil is present, bearing numerous finger-like processes. The body is pink to scarlet-red, with white sides. *Size:* Body length 4-8 in. (100-200 mm). *Habitat:* On sand; intertidal (in northern portions of its range) and subtidal to 300 ft. (91.4 m). *Distribution:* Central Gulf (distribution throughout the Gulf is not well known); from outer Baja California, north to Alaska; also Panama Bay, and Japan. *Remarks:* Deep water specimens from the Gulf of California have a deeper red color than animals collected from California and farther north. Occasionally captured by shrimp trawlers in large numbers.

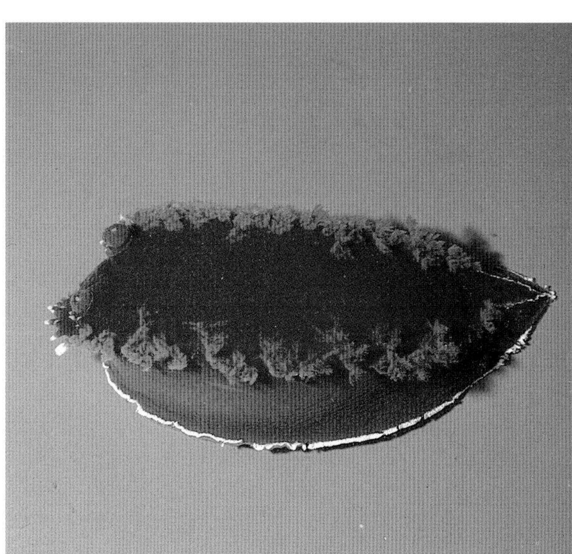

Hans Bertsch

148. ***Lomanotus stauberi***

Stauber's nudibranch

Description: The oral veil is semi-circular, with prominent, dorso-laterally projecting oral tentacles. Body margin has numerous branchial lobes. Body color is translucent white. *Size:* Body length under 0.5-1 in. (12-25 mm). *Habitat:* On the stinging hydroid *Lytocarpus* sp.; subtidal. *Distribution:* In the Gulf it has been recorded only from the Baja side of the central Gulf. It is also from Florida. *Remarks:* This nudibranch is well-camouflaged when on its hydroid prey. The 0.37-0.5 in. (9-12 mm) long egg mass is a thin, convoluted ribbon draped over the stalks of the host hydroid.

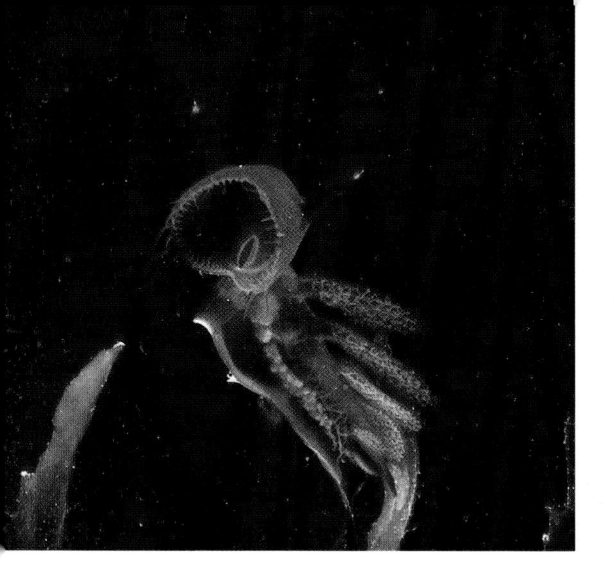

149. ***Melibe leonina***

Lion nudibranch

Description: The prominent cerata are laterally flattened. A large expandable oral hood, fringed with rows of small cirri, extends forward from the head. *Size:* Body length 2-4 in. (50-100 mm). *Habitat:* On floating *Sargassum* and other algae; subtidal and offshore to at least 50 ft. (15.2 m). It also occurs in *Macrocystis* kelp canopies along the Pacific coast of the U.S. and Canada. *Distribution:* Throughout the Gulf, outer coast of Baja California from Cabo San Lucas to Alaska. *Remarks:* This common nudibranch feeds on minute crustaceans, such as amphipods and copepods, by trapping them with its huge scoop-like oral hood. It is one of the widest ranging animals on our coast.

150. ***Histiomena convolvula***

Burrowing nudibranch

Description: With a bright pink foot bordered by three color bands: a thin white line, a broad central orange band, and an inner opalescent blue-white line. The dorsum bears convoluted ridges. *Size:* Body length 2-4 in. (50-100 mm). *Habitat:* In subtidal sandy substrate; to 150 ft. (45.7 m). *Distribution:* Throughout the Gulf (distribution outside the Gulf is unrecorded). *Remarks:* This sand-burrowing arminid feeds on gorgonian polyps of *Muricea* sp. and on the fleshy sea pen, *Ptilosarcus undulatus*. It is occasionally trawled by shrimpers.

Boris Innocenti

151. ***Dirona albolineata***

White lined dirona

Description: The cerata are flattened and lance-shaped. The whitish-gray body has prominent opaque white lines on the cerata and oral veil margin. *Size:* Body length 1-1.5 in. (25-38 mm). *Habitat:* Rocky intertidal and subtidal substrates; to 100 ft. (30.5 m). *Distribution:* Central and southern Gulf (distribution south of the Gulf is undetermined); also from southern California to British Columbia. *Remarks: Dirona albolineata* feeds on a wide variety of animals, including crustaceans, prosobranch snails, bryozoans, tunicates, and sponges.

152. ***Flabellina cynara***

Swimming cynara

Description: The long, salmon-orange cerata are tipped with white. There are purple streaks on posterior dorsal surface of the foot, on the rhinophores, and on the cephalic tentacles. *Size:* Body length 1.5-2 in. (38-50 mm). *Habitat:* On subtidal patch reefs and sandy bottoms; to at least 50 ft. (15.2 m). *Distribution:* Northern and central Gulf (distribution in the southern Gulf and outside the Gulf is uncertain). *Remarks:* This elegant eolid swims by using back-and-forth strokes of its whiplike cerata. The cerata gently curl forward in a loosely coordinated flowing movement, pause, and then whip backwards in a simultaneous wave. The rearward flipping of the cerata is the power stroke.

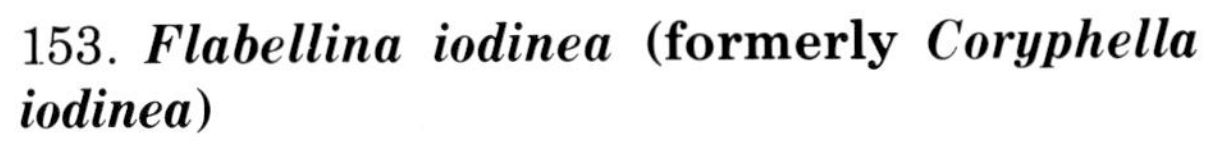

153. ***Flabellina iodinea* (formerly *Coryphella iodinea*)**

Spanish shawl

Description: The brilliant body color and long cerata quickly distinguish this species from all other Gulf nudibranchs. The intensity of the body coloration varies between individuals. *Size:* Body length 2-4 in. (50-100 mm). *Habitat:* Rocky intertidal and subtidal substrates; to 400 ft. (122 m). *Distribution:* Throughout the Gulf, outer Baja California Sur to British Columbia. *Remarks:* This species is seasonally and locally common. In the Gulf it is associated with colder water in winter and spring. In California it preys on the hydroid, *Eudendrium ramosum*.

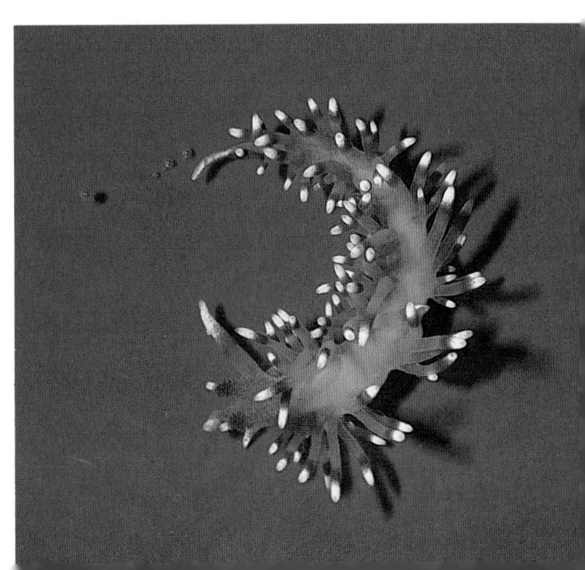

154. ***Flabellina telja***

Blue telja

Description: Body is bluish; tips of cephalic tentacles, rhinophores, and cerata are white. *Size:* Body length 0.5-0.75 in. (12-19 mm). *Habitat:* Under subtidal rocks and in crevices. *Distribution:* Throughout the Gulf (distribution outside the Gulf is unrecorded). *Remarks:* Additional taxonomic research is needed to properly identify this species as its status is questionable.

155. ***Bajaeolis bertschi***

Rainbow sea slug

Description: Numerous white dots give a frosted appearance to the head, sides of the body, and cerata. A purple ring encircles the middle of each oral tentacle. *Size:* Body length 0.5-0.75 in. (12-19 mm). *Habitat:* On hydroid colonies; subtidal. *Distribution:* Known from Bahía de los Angeles and Panama. *Remarks:* Its salmon-colored egg mass is occasionally found draped over its hydroid prey, *Eudendrium*.

156. ***Hermissenda crassicornis***

Hermissenda

Description: Ceratal cores are often briliant orange-brown, but many have more faded colors. The orange midline just behind the rhinophores is characteristic. *Size:* Body length 1.5-2 in. (38-50 mm). *Habitat:* Rocky intertidal and subtidal substrates. *Distribution:* Throughout the Gulf, and outer Baja California Sur from Punta Eugenia to Alaska. *Remarks:* It is a particularly common nudibranch along the coast of California and Oregon. In aquaria, these animals are non-specific scavengers.

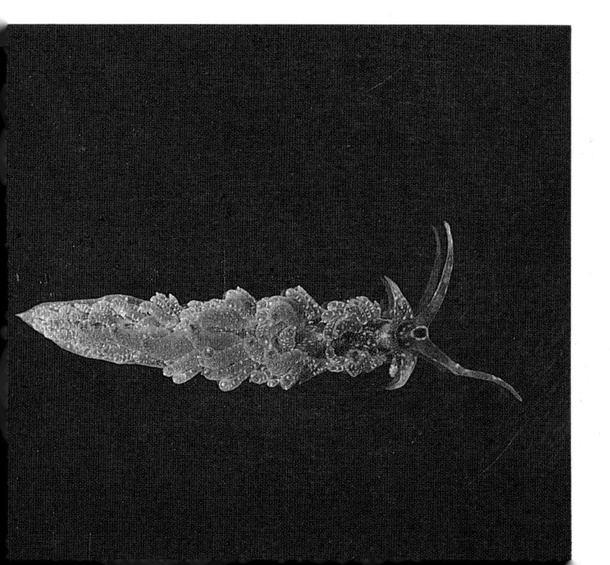

157. ***Baeolidia nodosa***

Knotty eolid

Description: The dorsum has concentric circles of various colors (yellow, pink, white, or blue-green bands). The spindle-shaped cerata may be slightly flattened and covered with small tubercles. *Size:* Body length 1-1.5 in. (25-38 mm). *Habitat:* On intertidal and subtidal rocky habitats. *Distribution:* Central and southern Gulf (distribution throughout the Gulf is not known); also recorded from the Mediterranean, West Atlantic, and Hawaii. *Remarks:* This species has been reported to feed on anemones of the genus *Bunodeopsis* in the Mediterranean. It is rare in the Gulf.

158. ***Spurilla neapolitana***

Naples' eolid

Description: The body color is variable from yellowish to pink, with opaque white spots on the dorsum and cerata. The color of cerata varies from green to brown, or dark gray. *Size:* Body length 1-2.5 in. (25-63 mm). *Habitat:* Rocky intertidal and subtidal substrates. *Distribution:* Central Gulf (Gulf distribution is poorly known); also recorded from the tropical Atlantic and the Hawaiian Islands. *Remarks:* The color variation is probably caused by its variable feeding habits. It feeds primarily on anemones. It is apparently rare in the Gulf.

Class Polyplacophora (Chitons)

159. ***Chiton virgulatus***

Virgulate chiton

Description: The dorsal surface is greenish on each plate and on the girdle, with light and dark alternating girdle bands. The inside of the shell plates are turquoise. *Size:* Body length 2-2.5 in. (50-63 mm). *Habitat:* On and under intertidal and subtidal rocks; to 20 ft. (6.1 m). *Distribution:* Throughout the Gulf, and outer Baja California Sur from Cabo San Lucas to Bahia Magdalena. *Remarks:* This is the most common chiton in the Gulf.

Class Cephalopoda (Octopuses, Squids)

160. ***Octopus bimaculatus***

Two spotted octopus

Description: Color is variable: mottled brown, gray, or rusty to yellow green, with two ocelli on the web between second and third arm. The skin is "wartier" than many other species. Cirri are present. *Size:* Total length (body plus the longest arm) 10-18 in. (250-450 mm). *Habitat:* Under intertidal and subtidal rocks and in crevices; to 150 ft. (45.7 m). *Distribution:* Throughout the Gulf to Panama; outer Baja California from Cabo San Lucas to southern California. *Remarks:* Like other octopods, this common species feeds largely on bivalve molluscs; octopus dens are usually surrounded by discarded, empty valves. This large species is taken commercially by divers throughout its range.

161. ***Octopus chierchiae***

Zebra octopus

Description: The unusual "zebra" color pattern distinguishes it from all other Gulf octopods. *Size:* Total length (body plus longest arm) 2.5-3.0 in. (63-75 mm). *Habitat:* Under intertidal and subtidal rocks, and inside empty gastropod shells; to 100 ft. (30.5 m). *Distribution:* Central Gulf to Panama. *Remarks:* This is a relatively rare species in the Gulf. Like many other octopods, it may bite if handled carelessly.

162. ***Octopus veligero***

Veligero octopus

Description: Superficially resembles *O. bimaculatus* but can be distinguished by its rusty orange funnel and mantle aperture and rosy undersides of arms with large white suckers. There are no ocelli. Dorsal surface varies from iridescent green, whitish, black, rusty red, and purple. The skin is smoother than *Octopus bimaculatus*. *Size:* Total length (body plus longest arm) 14-18 in. (350-450 mm). *Habitat:* Under intertidal and subtidal rocks and crevices; to 65 ft. (19.7 m). *Distribution:* Central Baja coast in the Gulf, and outer Baja California from Cabo San Lucas to San Juanico. *Remarks:* This octopus can be very abundant locally. Its prey include bivalves, cup-and-saucer limpets, snails (*Nerita*), and small crabs.

163. ***Octopus digueti***

Pigmy octopus

Description: The color at rest is tan, but by controlling its pigment cells a multitude of color patterns can be achieved. It is cryptic against a sand background, but flushes red when excited. The skin is smooth with transient papillae. *Size:* Total length (body plus longest arm) 8-10 in. (200-250 mm). *Habitat:* Restricted to sandy and muddy substrates, in empty gastropod and bivalve shells, as well as discarded bottles and cans. Intertidal and subtidal to 120 ft. (36.6 m). *Distribution:* Throughout the Gulf to at least Mazatlán. *Remarks:* If handled carelessly the bite of this species is particularly irritating. It preys upon brachyuran crabs and hermit crabs. This octopus is locally common, particularly in northern Sonora.

164. ***Octopus fitchi***

Fitch's octopus

Description: Overall color is normally brownish-red with a purple undertone, but can become pale gray or beige when excited. Golden spots may be present near arms one and two. The skin has several papillae and the body is globular. *Size:* Total length (body plus longest arm) 3.5-5 in. (87-125 mm). *Habitat:* Under intertidal and subtidal rocks, and on coquina limestone. *Distribution:* Throughout the Gulf (distribution outside the Gulf is unrecorded). *Remarks:* The eggs are 0.25 in. (6 mm) long, producing planktonic young. This uncommon species can inflict painful bites that may be venomous. The prey include crabs and other crustaceans.

165. ***Octopus alecto***

Longarm octopus

Description: The arms are extremely long relative to body length. No ocelli are present. The skin is generally smooth; supraocular cirri may be present. Overall color is superficially similar to *O. fitchi,* but bright light green iridiocytes may also be discernable. *Size:* Total length (body plus longest arm) 5-5.25 in. (125-137 mm). *Habitat:* On and under intertidal and subtidal rocks and coral reefs. *Distribution:* Throughout the Gulf to Nayarit, Mexico. *Remarks:* In captivity this uncommon species preys on porcelain crabs.

166. ***Dosidicus gigas***

Jumbo or Humboldt squid

Description: The body is rusty to dark brown. The posterior fins are well developed, the suckers are toothed, and the tentacles are webbed. Males are smaller than females. *Size:* Total length (body plus longest arm) 3-13 ft. (1-4 m). *Habitat:* Inshore and offshore; epipelagic to several hundred meters. *Distribution:* Throughout the Gulf to Chile, outer coast of Baja California from Cabo San Lucas to northern California; also reported from Australia. *Remarks:* Large numbers periodically migrate into the Gulf. The last run may have appeared in the winter of 1977. Periodic inshore migrations off the southern California coast are reported to occur every 35-40 years. These large squids feed on crustaceans, fishes, and other squids.

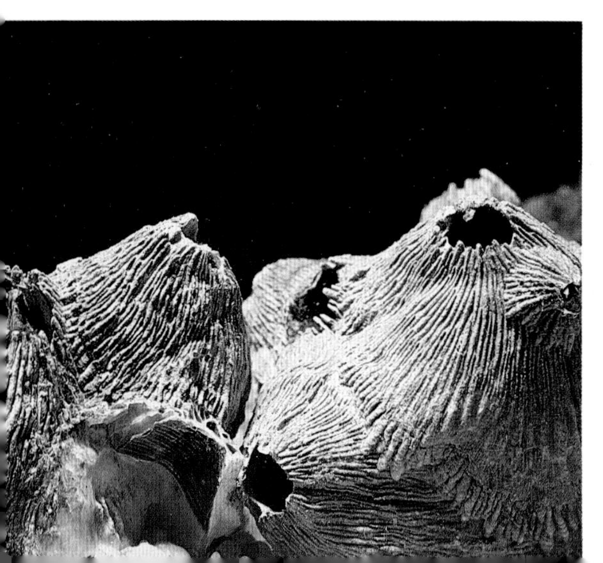

167. ***Argonauta cornuta***

Paper nautilus

Description: The delicate ribbed shell is milk-white to whitish tan with a fine granular texture. The basal keel is relatively broad and the two axial extensions are tinged with purplish brown to purplish black. *Size:* Shell width 2-3 in. (50-75 mm); shell length 3-3.75 in. (75-95 mm). *Habitat:* Pelagic, usually swimming just below the surface. *Distribution:* Throughout the Gulf to Panama. *Remarks:* Although rarely seen, this species must be fairly common offshore as large numbers of empty shells are occasionally found washed ashore after heavy storms or hurricanes. Only females secrete the thin, calcareous shells into which they deposit their eggs; the females are not attached to these shells. The males are considerably smaller and shell-less.

PHYLUM ARTHROPODA
Class Crustacea
(Barnacles, Shrimps, Lobsters, Crabs)

168. ***Lepas anserifera*** **Goose barnacle**

Description: The five plates protecting the body are furrowed. The stalks are shorter than in most other *Lepas* species. *Size:* Stalk length 1-2 in. (25-50 mm); body width 0.25-0.5 in. (6-12 mm). *Habitat:* Attached to drifting material such as wood, styrofoam, floats; also occasionally on pilings, and some hard-shelled animals. *Distribution:* Throughout the Gulf, cosmopolitan. *Remarks:* Large clusters often encrust floating objects such as wooden planks, bottles, and boats. Some crustaceans and sea turtles are occasionally infested with encrustations of this goose barnacle.

169. ***Tetraclita stalactifera*** **Volcano barnacle**

Description: The large, rough-sided shell is normally light gray to greenish or bluish gray. This non-stalked barnacle has a shell wall of four fused plates. *Size:* Base diameter 1-2 in. (25-50 mm). *Habitat:* On exposed rocks, in higher mid intertidal zone. *Distribution:* Throughout the Gulf to central west Mexico. *Remarks:* This is one of the largest and most common Gulf barnacles. Several subspecies are known from the Pacific coast of Mexico and the western Atlantic.

170. ***Chthamalus anisopoma* (formerly *Chthamalus fissus*)**

Acorn barnacle

Description: The plates vary from whitish to gray. The shell is composed of six plates; the end plates are overlapped by the larger side plates. *Size:* Diameter 0.25-0.5 in. (6-12 mm). *Habitat:* Attached to rocks in high to mid-intertidal zones. *Distribution:* Throughout the Gulf (may not occur south of the Gulf), along the outer coast of Baja California from Cabo San Lucasnal to central California. *Remarks:* This is the most common intertidal barnacle in the Gulf. It is preyed on by the Gulf sun star, *Heliaster kubiniji,* and the thorn shell, *Acanthina angelica.*

171. ***Lironeca vulgaris***

Fishgill isopod

Description: The legs have long pointed tips for clinging to the host fish. The symmetrical body is ovate and flattened, with an abdomen that tapers smoothly from the thorax. The head is slightly sunken into the first of the thoracic segments. Males are smaller than females. *Size:* Total body length 0.5-1.75 in. (12-44 mm). *Habitat:* Parasitic on a variety of marine fishes; reported on at least 30 species of host fishes. From the surface to 950 ft. (289.4 m). *Distribution:* Throughout the Gulf to Colombia, and outer Baja California from Cabo San Lucas to Oregon. *Remarks:* Both males and females are usually attached to the gills of host fishes, where they feed on blood. Females occasionally migrate into the host's mouth. This and the following species of isopod fish parasites (often called "fish lice") are sequential hermaphrodites.

172. ***Renocila thresherorum***

Fishscale isopod

Description: Body broadly ovate with scattered brown to purple spots on back; all legs with long, pointed, spine-like tips for clinging to host fish. Males are smaller than females. *Size:* Body length 0.5-1.25 in. (12-32 mm). *Habitat:* Parasitic on the body surface of various fishes; subtidal to below 200 ft. (61 m). *Distribution:* Central and southern Gulf (distribution outside the Gulf is unrecorded). *Remarks:* This isopod is usually found attached near the dorsal fin. They are most commonly found on fishes inhabiting wave-swept rocky shores.

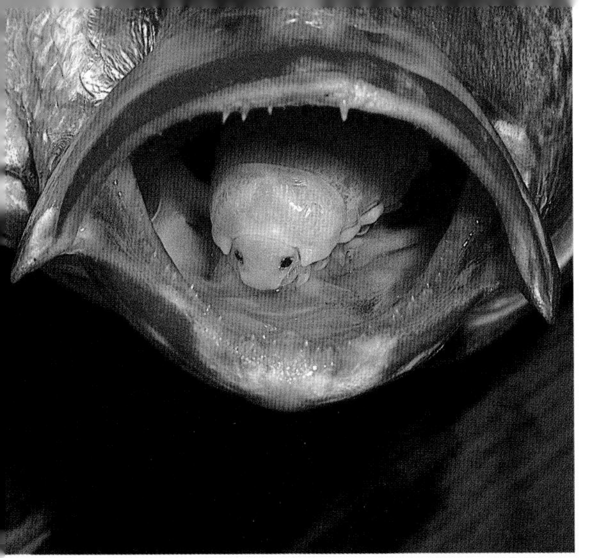

Mathew Gilligan

173. *Cymothoa exigua*

Tounge-eating isopod

Description: The body is slightly asymmetrical with a small head and legs ending in long, hook-like tips for clinging to host fishes. Males are smaller than females. *Size:* Body length 0.25-1.25 in. (6-32 mm). *Habitat:* Parasitic in mouths and gills of numerous marine fishes, including snappers, grunts, mullets, and corvinas; subtidal to at least 90 ft. (27.4 m). *Distribution:* Throughout the Gulf to Ecuador. *Remarks:* These isopods feed on the tongue of host fishes. Researchers hypothesize that tongue-eating isopods act as functional tongue replacements while clinging to the remaining tongue stub. In this manner a fish may continue feeding with the use of the "isopod-tongue."

174. *Ligia occidentalis*

Rock isopod

Description: The body is flattened dorsoventrally with two long, forked tail appendages and a pair of very long antennae. *Size:* Body length 0.75-1.25 in. (19-32 mm). *Habitat:* On rocky shores, near the water, but always out of the water line; often on dried algae drifts. *Distribution:* Throughout the Gulf to Ecuador; outer Baja California from Cabo San Lucas to central California. *Remarks:* This ubiquitous scavenger feeds on beach debris. It has a diurnal rhythm of color change: pale at night and dark during the day.

175. *Squilla mantoidea*

Barbedclaw mantis shrimp

Description: The uropods are bicolored, dark brown or gray and tipped with yellow. There are light brown bands on the posterior half of each abdominal segment. The claws are barbed. *Size:* Body length 4-10 in. (100-250 mm). *Habitat:* On sand or mud bottoms; subtidal at depths of 80-200 ft. (24.6-61 m). *Distribution:* Northern Gulf to Peru. *Remarks:* Once thought to be a rare species, *Squilla mantoidea* is regularly taken by shrimp trawlers; it is not as common as *Squilla tiburonensis*.

176. ***Squilla tiburonensis***

Spearing mantis shrimp

Description: The uropods and telson are tipped with black and edged with pale yellow. There are two squarish spots on the abdomen. The claws are barbed. *Size:* Body length 3-6 in. (75-150 mm). *Habitat:* On sand or mud; subtidal at depths of 15-250 ft. (4.6-76.2 m). *Distribution:* Throughout the Gulf (endemic to the Gulf). *Remarks:* This abundant species is usually collected in shrimp trawls. It lives in sand burrows or small depressions, catching prey by spearing small fishes and other soft bodied animals with its barbed claws.

177. ***Parasquilla similis***

Brown mantis shrimp

Description: There are regularly spaced, small whitish blotches on each abdominal segment. Small light dishes extend centrally on each thoracic and abdominal segment. The claws bear teeth. *Size:* Body length 4-8 in. (100-200 mm). *Habitat:* On sand, offshore at depths of 100-150 ft. (30.5-45.7 m). *Distribution:* Central Gulf to Peru. *Remarks:* This rare species is occasionally taken by shrimp trawlers. It preys on small fishes and other soft bodied animals.

178. ***Gonodactylus oerstedii***

Reef mantis shrimp

Description: Body is usually a mottled green or brown. Marginal teeth of the telson are prominent. Claws are without teeth. *Size:* Body length 1.25-1.5 in. (32-38 mm). *Habitat:* This mantis shrimp burrows under intertidal and subtidal rocks; to 40 ft. (12.2 m). *Distribution:* Throughout the Gulf to Ecuador; also the western Atlantic. *Remarks:* This is one of the most common rocky shore mantis shrimps in the Gulf.

179. ***Hemisquilla ensigera californiensis***

Peacock mantis shrimp

Description: The brilliant tail of this species distinguishes it from all other Gulf mantis shrimps. The claws are sharp-tipped but are not barbed. *Size:* Body length 4-12 in. (100-300 mm). *Habitat:* On sand; offshore at depths of 100-300 ft. (30.5-91.4 m). *Distribution:* Central Gulf to Panama, also reported from southern California. *Remarks:* This relatively common species lives in deep burrows but forages on sand, feeding on shrimp, bivalves, and snails at night. Specimens should be handled with gloves because their sharp claws can inflict serious lacerations. Large numbers of these mantis shrimps are regularly captured in shrimp trawls.

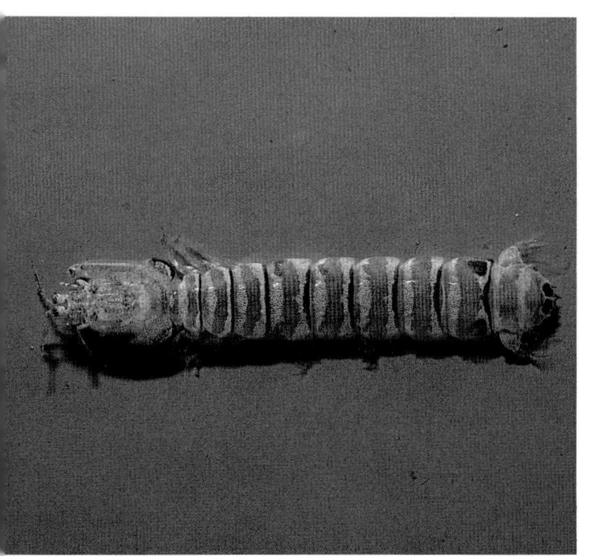

180. ***Acanthosquilla digueti***

Pinkback mantis shrimp

Description: There are two lateral black spots on the sixth abdominal segment and on the tail. There is a thin dark band on the posterior edge of each abdominal segment. *Size:* Body length 0.75-2 in. (19-50 mm). *Habitat:* On sand or mud; intertidal and subtidal to 100 ft. (30.5 m). *Distribution:* Throughout the Gulf to Panama, and throughout the tropical western Atlantic. *Remarks:* This small mantid lives along with the polychaete worm, *Lepidasthenia digueti,* both being commensals in the burrow of the acorn worm, *Balanoglossus.*

181. ***Microprosthema emmiltum***

Elusive shrimp

Description: The overall body is white, except for the red antennae and legs. The pair of claws are particularly large. *Size:* Body length 0.37-0.5 in. (9-12 mm). *Habitat:* Under subtidal rocks; to 10 ft. (3 m). *Distribution:* In the Gulf it is known only from Cabo San Lucas; it is also known outside the Gulf from Panama and the Galápagos. *Remarks:* This new species has been collected only once in the Gulf, at Cabo San Lucas.

182. ***Odontozona rubra***

Ruby shrimp

Description: The body is translucent with red bands on the abdomen and red lines on carapace. The clawed legs are particularly long. *Size:* Body length 0.5-1 in. (12-25 mm). *Habitat:* Offshore islands in caves and crevices; subtidal at depths of 15-30 ft. (4.6-9 m). *Distribution:* Central and southern Gulf (may be a Gulf endemic). *Remarks:* This newly described species is rare and has been observed to clean fishes of parasites. It is related to the coral shrimp, *Stenopus*.

183. ***Veleronia laevifrons***

Sea fan shrimp

Description: The body color is variable and will normally mimic the host's body color. *Size:* Body length 0.25-0.37 in. (6-9 mm). *Habitat:* This shrimp lives as a commensal on at least two species of gorgonians, *Eugorgia aurantica* and *E. ampla;* subtidal at depths of 10-60 ft. (3-18.3 m). *Distribution:* Central Gulf to Galápagos. *Remarks:* A single gorgonian may have 3 to 10 individual shrimps clinging to its branches.

184. ***Palaemon ritteri***

Tidepool shrimp

Description: This transparent shrimp has dark stripes with small reddish spots on the tail and red banding on the legs. *Size:* Body length 0.75-1.5 in. (19-38 mm). *Habitat:* Shallow intertidal and subtidal rocky reefs; to 10 ft. (3 m). *Distribution:* Throughout the Gulf to Peru; also southern California. *Remarks:* This ubiquitous shrimp lives under rocks and in crevices during the day but forages on top of rocks and sand at night. This shrimp may be a cleaner. The pictured specimen is a female (note eggs under abdomen).

185. ***Brachycarpus biunguiculatus***

Clawed shrimp

Description: The body is pink to reddish with a blue-green mottling. The pair of clawed legs are particularly long. *Size:* Body length 1-2 in. (25-50 mm). *Habitat:* In subtidal cliff crevices; at depths of 10-50 ft. (3-15.2 m). *Distribution:* Central Gulf to Ecuador; circumtropical. *Remarks:* This uncommon species has been observed cleaning fishes.

186. ***Periclimenes lucasi***

Lucas' cleaner shrimp

Description: Easily recognized by its distinct bright coloration, particularly the alternating blue-and-yellow banding on the legs. *Size:* Body length 0.25-1 in. (6-25 mm). *Habitat:* On sand, mud, or rubble. It is associated with certain species of cnidarians, particularly tube anemones, *Pachycerianthus*. Subtidal at depths of 35-150 ft. (10.7-45.7 m). *Distribution:* Central Gulf to Panama. *Remarks:* this species is a cleaner of fishes. When approached by divers it often performs a ritualistic "dance" by rocking its body from side to side. Unlike the Atlantic species of *Periclimenes,* which readily mingle among anemone tentacles, Lucas' cleaner shrimp lives only at the base of the tube anemone, avoiding the long tentacles.

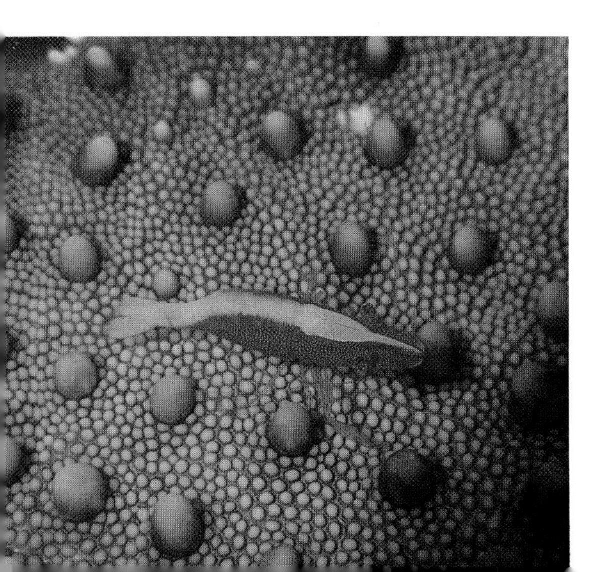

187. ***Periclimenes soror***

Sea star shrimp

Description: The body color is variable from white to rusty-brown or yellow, often bicolored. This shrimp normally mimics its host's coloration. *Size:* Body length 0.25-0.5 in. (6-12 mm). *Habitat:* Lives commensally on the oral (bottom) side of sea stars; at depths of 40-100 ft. (12.2-30.5 m). *Distribution:* Central Gulf to Colombia; circumtropical. *Remarks:* Recently discovered from the Gulf, this locally common shrimp is associated with at least 5 species of sea stars in the Gulf (16 worldwide) including *Pentaceraster cumingi* (fomerly *Oreaster occidentalis*), *Astrometis sertulifera, Nidorellia armata, Mithrodia bradleyi,* and *Acanthaster ellisii.*

188. ***Harpiliopsis depressa***

Coral shrimp

Description: Pinkish tan to pale green with faint darker bands on the abdomen. The long pair of clawed legs are also banded. *Size:* Body length 1-1.75 in. (25-44 mm). *Habitat:* In deep crevices and coral heads (*Pocillopora elegans*); at depths of 15-40 ft. (4.6-12.2 m). *Distribution:* Southern Gulf to Ecuador; circumtropical. *Remarks:* It is common inside nearly all heads of the elegant coral, *Pocillopora elegans*.

189. ***Pontonia pinnae***

Pearl shrimp

Description: The body is semi-translucent whitish pink. Females are larger than males, with distinctly larger abdomens. *Size:* Body length 1.75-1.5 in. (19-38 mm). *Habitat:* Normally living as commensals inside the pearl oyster, *Pinctada mazatlanica,* and pen shells, *Pinna rugosa* and *Atrina tuberculosa;* at depths of 10-100 ft. (3-30.5 m). *Distribution:* Throughout the Gulf to Panama. *Remarks:* Usually one pair, a male and a female, are found together inside each pen or pearl shell.

190. ***Palaemonella holmesi***

Longclaw shrimp

Description: The body is transparent to pale brown, with a pair of long clawed legs. *Size:* Body length 0.25-1 in. (6-25 mm). *Habitat:* In rubble and crevices; at a depth of 6-300 ft. (2-91.4 m). *Distribution:* Throughout the Gulf to Ecuador; outer coast of Baja California from Cabo San Lucas to southern California. *Remarks:* Individuals of this species are larger in California, reaching over 2 in. (50 mm) in length.

191. ***Thor algicola* (formerly *Thor paschalis*)**

Pigmy shrimp

Description: The body is transparent, with a network of white and blue spots. *Size:* Body length 0.25-0.37 in. (6-9 mm). *Habitat:* Under rocks and in clumps of *Sargassum* algae attached to rocks and jetties; intertidal to 60 ft. (18.3 m). *Distribution:* Throughout the Gulf to Panama. *Remarks:* Once thought to be rare, this species is actually quite common.

192. ***Gnathophyllum panamense***

Cortez barrel shrimp

Description: The thick body is spotted and the clawed legs are whitish. Females are larger than males. *Size:* Body length 0.5-1 in. (12-25 mm). *Habitat:* Under intertidal and subtidal rocks and in crevices; to 90 ft. (27.4 m). *Distribution:* Throughout the Gulf to Ecuador. *Remarks:* This species will remove parasites from infested fishes. It also feeds on the tube feet of certain sea urchins.

193. ***Alpheus sulcatus***

Furrowed shrimp

Description: Body color varies from pale green to orange or deep brown. The claws usually have small blue spots and the sides of the carapace bear small white dots. *Size:* Body length 1.25-3 in. (32-75 mm). *Habitat:* Under intertidal and subtidal rocks; to 30 ft. (9.1 m). *Distribution:* Throughout the Gulf to Peru; circumtropical. *Remarks:* This common species is one of the largest of the Gulf pistol shrimps. It has been traditionally called *Alpheus californiensis*. However, researchers have shown that *A. californiensis* only occurs from southern California to Bahía Magdalena, Baja California Sur, and not in the Gulf.

194. ***Alpheus paracrinitus***

Barred pistol shrimp

Description: The body is translucent with six distinct maroon to orange bars on the abdomen and thorax. *Size:* Body length 0.5-1.25 in. (12-32 mm). *Habitat:* Under intertidal and subtidal rocks and rubble; to 60 ft. (18.3 m). *Distribution:* Throughout the Gulf to Ecuador; circumtropical. *Remarks:* This species is uncommon in the Gulf.

195. ***Alpheus grahami***

Blueleg pistol shrimp

Description: Similar to other Gulf pistol shrimps, but distinguished by its bluish legs and reddish claws with violet tips. *Size:* Body length 1-1.5 in. (25-38 mm). *Habitat:* Under intertidal rocks, in crevices, and under coral slabs; to 150 ft. (46 m). *Distribution:* Central Gulf to Colombia. *Remarks:* This species is uncommon in the Gulf.

196. ***Alpheus lottini***

Coral pistol shrimp

Description: This is the most colorful of all Gulf shrimps. It can be readily identified by its distinctive combination of spotted claws and black dorsal pattern. *Size:* Body length 1.25-2 in. (32-50 mm). *Habitat:* Among branches of the coral, *Pocillopora* spp.; subtidally to 50 ft. (15.2 m). *Distribution:* Lower Gulf to Ecuador; Indo-West Pacific. *Remarks:* Several individuals may inhabit a single coral head. It is common within its range.

197. ***Synalpheus digueti***

Sponge pistol shrimp

Description: Body is generally translucent green. This species can be distinguished from other Gulf pistol shrimp by its reddish claws with black tips. Females with eggs are dull yellowish pink. *Size:* Body length 0.5-1 in. (12-25 mm). *Habitat:* Unlike most pistol shrimps which burrow under rocks or rubble, this small species is found inside sponges that encrust reef ledges, rocky outcroppings, and jetties. Paired individuals may also live in empty polychaete tubes; intertidal and subtidal to 50 ft. (15.2 m). *Distribution:* Throughout the Gulf to Ecuador. *Remarks:* As in other pistol shrimp, this common species produces a metallic or glass-like snapping retort by snapping the fingers of the large claw together forcefully.

198. ***Lysmata californica***

Peppermint cleaner shrimp

Description: Body is normally a semi-transparent pink to red with reddish transverse stripes. *Size:* Body length 1.25-2 in. (32-50 mm). *Habitat:* Under intertidal and subtidal rocks, in crevices, and in caves; to 30 ft. (9.1 m). *Distribution:* Throughout the Gulf to Panama, outer Baja California from Cabo San Lucas to northern California. *Remarks:* This nocturnal shrimp is most common in the upper Gulf and in southern California. Like most other *Lysmata,* this species is known to clean fishes of external parasites, dead skin, and other debris.

199. ***Lysmata intermedia***

Intermediate cleaner shrimp

Description: The body is pinkish to red, with several light silver stripes running along the sides; there are white dots within the stripes. *Size:* Body length 1.25-2 in. (32-50 mm). *Habitat:* Under intertidal and subtidal rocks, in crevices, and caves; to 60 ft. (18.3 m). *Distribution:* Central Gulf to Peru; also the Caribbean. *Remarks:* Like most other *Lysmata,* this species is a fish cleaner.

200. ***Lysmata galapagensis***

Banded cleaner shrimp

Description: The body is red with several distinct white bands on the dorsal part of the abdomen. *Size:* Body length 1-1.25 in. (25-32 mm). *Habitat:* Under intertidal and subtidal rocks, in crevices; to 60 ft. (18.3 m). *Distribution:* Central Gulf to Ecuador. *Remarks:* This species is uncommon in the Gulf and is found mostly around offshore islands. It is a known fish cleaner.

201. ***Penaeus californiensis***

Brown or khaki shrimp

Description: The body is tan to reddish brown. A well-developed groove runs along both sides of the rostrum. A dorsolateral groove occurs on the last abdominal segment. Females are larger than males. *Size:* Body length 6-10 in. (150-250 mm). *Habitat:* Offshore on sand, at depths of 60-300 ft. (18.3-91.4 m). *Distribution:* Throughout the Gulf, north to southern California and south to Peru. *Remarks:* This is one of the most important of the commercially harvested shrimps. It comprises about 60% of the Pacific Mexican shrimp fishery. Juveniles do not occur in estuaries. Adults live 3 to 4 years.

202. ***Penaeus stylirostris***

Blue shrimp

Description: The body is white to pale blue. A lateral groove is absent, or barely visible, along both sides of the rostrum. Females are larger than males. *Size:* Body length 5-11 in. (125-275 mm). *Habitat:* Offshore in sand at depths of 50-250 ft. (15.2-76.2 m). *Distribution:* Throughout the Gulf to Peru. *Remarks:* This is one of the commercially harvested species. Juveniles occur in estuaries. In captivity females can grow to 8-10 in. (200-250 mm) in 12-14 months, and live 3 to 4 years.

203. ***Sicyonia penicillata***

Target shrimp

Description: The eye-spot markings on the side of the carapace are distinct, but the overall color is variable (three color variations). *Size:* Body length 0.5-2 in. (12-50 mm). *Habitat:* Offshore on sand at depths of 50-300 ft. (15.2-91.4 m). *Distribution:* Throughout the Gulf (distribution outside the Gulf is unrecorded). *Remarks:* This is the most common species of *Sicyonia* in the Gulf and is now commercially harvested by shrimp fishermen.

204. ***Sicyonia aliaffinis***

Bullseye shrimp

Description: The rostrum has two dorsal teeth. The posterior carapace tooth is elevated. *Size:* Body length 1.25-2 in. (32-50 mm). *Habitat:* Offshore on sand at depths of 100-300 ft. (30.5-91.4 m). *Distribution:* Central Gulf to Peru. *Remarks:* This uncommon shrimp is occasionally trawled by shrimpers.

205. ***Sicyonia disedwardsi***

Goldring shrimp

Description: The eyespot markings on the sides of the carapace have a gold ring around a purplish center. *Size:* Body length 1-1.5 in. (25-38 mm). *Habitat:* Offshore on sand at depths to 100 ft. (30.5 m). *Distribution:* Central and lower Gulf (distribution outside the Gulf is unrecorded). *Remarks:* This uncommon species is occasionally trawled by shrimpers.

206. ***Axius vivesi***

Hairyclaw ghost shrimp

Description: This large shrimp resembles a crayfish. The large claws are reddish and bristled. *Size:* Body length 3-5 in. (75-125 mm). *Habitat:* In limestone, sand, or rubble burrows; low intertidal and subtidal to 150 ft. (46 m). *Distribution:* Throughout the Gulf (distribution outside the Gulf is unrecorded). *Remarks:* This common, secretive shrimp is usually seen 1 or 2 inches inside the entrance of its burrow.

207. ***Panulirus inflatus***

Pinto spiny lobster

Description: The body is dark brownish-gray, with orange carapace spines. The abdomen has numerous white spots. *Size:* Body length 10-20 in. (25-50 cm). *Habitat:* In rock crevices and caves at depths of 5-250 ft. (1.5-76.2 m). *Distribution:* Central Gulf to the Gulf of Tehuantepec, southern Mexico, and the outer coast of Baja California north to Bahia Magdalena. It is rare north of Bahía Magdalena; two records have been reported from San Diego, California (1961 and 1962). *Remarks:* This is the most common species in the Gulf, but its abundance is declining due to overfishing by sport and commercial divers.

208. ***Panulirus gracilis***

Blue spiny lobster

Description: The body is dark greenish or bluish-gray, with green or blue carapace spines. The abdomen lacks white spots. *Size:* Body length 10-12 in. (25-30 cm). *Habitat:* In rock crevices and caves; subtidal at depths of 10-150 ft. (3-46 m). *Distribution:* Central Gulf to Peru. *Remarks:* This species is uncommon in the Gulf, becoming more abundant in southern Mexico where *Panulirus inflatus* seldom occurs. In the Gulf *P. gracilis* appears to be associated with deep water patch reefs.

Don Thomson

Gary Cotter

209. ***Panulirus interruptus***

California spiny lobster

Description: Overall body is reddish. The carapace spines are blunt and fewer in number than in other Gulf spiny lobsters. *Size:* Body length 12-28 in. (30-70 cm). *Habitat:* In rock crevices and caves; subtidal at depths of 5-250 ft. (1.5-76.2 m). *Distribution:* Upper Gulf (Baja) only at Bahía de Los Angeles, and outer Baja California to southern California. Rarely to northern California. Does not occur south of the Gulf. *Remarks:* Although uncommon in the lower Gulf, small populations have been regularly reported from canyons at Los Frailes, Gorda Banks, and Cabo San Lucas.

210. ***Evibacus princeps***

Panamic slipper lobster

Description: The body is a yellowish-brown to dark reddish-brown with a hard, flat carapace. Overall shape of the entire lobster is round, somewhat resembling a horseshoe crab. *Size:* Body length 10-14 in. (25-35 cm). *Habitat:* Offshore on sand at depths of 100-500 ft. (30.5-152.4 m). *Distribution:* Throughout the Gulf to Peru. *Remarks:* This uncommon lobster is nocturnal and occasionally trawled by shrimpers; it is rarely sold in local markets.

211. ***Scyllarides astori***

Rock slipper lobster

Description: The body is tan to rusty brown, with a purple edge on the anterior of the carapace. The body shape is more oblong than that of *Evibacus princeps*. *Size:* Body length 10-18 in. (25-45 cm) *Habitat:* Under rocky ledges or in crevices, usually close to sand; at depths of 40-300 ft. (12.2-91.4 m) *Distribution:* Throughout the Gulf to Ecuador *Remarks:* This lobster is relatively common locally. Unlike *Evibacus princeps,* it appears to be associated with rocky substrates around offshore islands. It is preyed upon by large reef fishes such as the Gulf grouper, *Mycteroperca jordani,* and the jewfish, *Epinephelus itajara.*

212. ***Euceramus transversilineatus***

Commensal porcelain crab

Description: The pinkish body is marked with red or brown striations. Long bristles cover the claws. *Size:* Carapace length 0.25-0.75 in. (6-19 mm). *Habitat:* Intertidal and on subtidal rocks; to 120 ft. (36.6 m). *Distribution:* Throughout the Gulf to Panama. *Remarks:* This uncommon crab is occasionally found attached to gorgonians, tube anemones, and sea pens.

213. ***Petrolisthes marginatus***

Striped porcelain crab

Description: The body is pale orange with red striations on the carapace and claws. *Size:* Carapace length 0.5-0.75 in. (12-19 mm). *Habitat:* Among branches of the coral, *Pocillopora elegans;* low intertidal and subtidal to 60 ft. (18.3 m). *Distribution:* Central Gulf to Ecuador, including Islas Tres Marias, Clipperton, and Revillagigedo; also reported from the eastern Caribbean. *Remarks:* This species appears to be a regular commensal with *Pocillopora elegans*.

214. ***Porcellana* sp. (species is undertermined)**

Scallop porcelain crab

Description: The body color of this small crab mimics red-striped barnacles. *Size:* Carapace length 0.5 in. (12 mm). *Habitat:* Specimens have been collected on the surface of scallop shells, *Argopecten circularis. Distribution:* Central and southern Gulf (distribution outside the Gulf is undetermined). *Remarks:* When clinging to the valve of a scallop, this rare commensal crab is difficult to see, closely resembling the attached barnacles on the shell.

215. ***Pleuroncodes planipes***

Pelagic red crab, lobster krill

Description: The body is brilliant reddish orange to brownish red. The clawed legs are long and slender. *Size:* Carapace length 1.5-2 in. (38-50 mm). *Habitat:* Offshore; epipelagic to 1800 ft. (600 m). *Distribution:* Throughout the Gulf to at least central Mexico, the outer coast of Baja California from Cabo San Lucas to north of San Francisco, California. *Remarks:* Occasionally enormous numbers of this species wash ashore. A similar species, *Pleuroncodes monodon,* is often trawled by shrimpers and is currently captured commercially in large quantities off the coasts of Chile and El Salvador. Only the tail meat is used and marketed. Because this crab contains the red pigment, carotenoid, it is fed to commercially raised salmon to retain the reddish-pink flesh color.

216. ***Aniculus elegans***

Elegant hermit crab

Description: The legs and claws are pink and irregularly banded with red. The tips of the claws are bright red and grooved. The hairs on the body and appendages are tan, tipped with white. *Size:* Carapace length 1.5-3 in. (38-75 mm). *Habitat:* On sand, rubble, and among rocks; low intertidal and subtidal 150 ft. (46 m). *Distribution:* Central Gulf to Ecuador. *Remarks:* This common species is one of the largest Gulf hermit crabs. It inhabits the shells of large gastropods like *Murex, Strombus,* and *Pleuroploca.* Occasionally trawled by shrimpers.

217. ***Clibanarius digueti***

Bluespotted hermit crab

Description: The leg color is variable from reddish brown to gray, or olive, with bluish spots. The antennae and antennules are brilliant red. Both claws are hairy and equal in size. *Size:* Carapace length 0.25-0.5 in. (6-12 mm). *Habitat:* High and mid-intertidal zone. *Distribution:* Throughout the Gulf. *Remarks:* This is the most common hermit crab in the Gulf. Several hundred individuals may be found clustered under rocks at low tide. They scavenge and filter feed. The parasitic isopod, *Pseudione* sp. has been reported by R.C. Brusca to be associated with this species.

218. ***Dardanus sinistripes***

Anemone carrier hermit crab

Description: The overall body is pinkish-gray to orange-brown. Tufts of hair partially cover the legs and claws. The eyestalks are banded, but faintly so in some individuals. The left claw is considerably larger than the right. *Size:* Carapace length 0.5-1 in. (12-25 mm). *Habitat:* On sand and rubble substrates; at depths of 6-330 ft. (1.8-100.5 m). *Distribution:* Central Gulf to Peru. *Remarks:* Color variants of this relatively common hermit crab may represent another species. It commonly carries one or more anemones (*Calliactis*) on its shell.

219. ***Paguristes sanguinimanus***

Blue-eye hermit crab

Description: The overall body is orange, as opposed to dark red in the similar *Paguristes digueti.* There are three strong spines on the fifth segment (from the base) of the cheliped in *P. digueti* that are lacking in *P. sanguinimanus.* Both species have bright blue eyes. *Size:* Carapace length 0.25-0.5 in. (6-12 mm). *Habitat:* On intertidal and subtidal sand flats and patch reefs. *Distribution:* Throughout the Gulf. *Remarks:* Large clusters of this hermit crab are occasionally found on low intertidal and subtidal patch reefs. It is preyed upon by octopods.

220. ***Calcinus californiensis***

Redleg hermit crab

Description: The walking legs are solid orange or red; the chelae are greenish-black with reddish margins. *Size:* Carapace length 0.5-0.75 in. (12-19 mm). *Habitat:* Rocky substrates in the low intertidal and shallow subtidal zones. In the southern Gulf it is often found in coral heads (*Pocillopora*). *Distribution:* Central Gulf to Acapulco. *Remarks:* It is not unusual to find large aggregations of this species.

221. ***Petrochirus californiensis***

Giant hermit crab

Description: The overall body is rusty brown, with pronounced knobs on the claws. *Size:* Carapace length 1-3 in. (25-75 mm). *Habitat:* Rock and sand substrates; at depths of 10-100 ft. (3-30.5 m). *Distribution:* Throughout the Gulf to Ecuador. *Remarks:* This common species is the largest of the Gulf hermit crabs. It occupies various empty gastropod shells, particularly *Murex, Pleuroploca,* and *Strombus*. Commensal porcelain crabs and some species of worms are occasionally associated with this hermit crab.

222. ***Manucomplanus varians* (formerly *Pylopagurus varians*)**

Staghorn hermit crab

Description: The large claw acts as a "trap-door" to seal the entrance of the staghorned hydrocoral "shell" that is selected exclusively by the crab. The opening of the hydrocoral "shell," *Janaria mirabilis,* is apparently kept manicured by the crab. *Size:* Carapace length 0.25 in. (6 mm). *Habitat:* On sand, often close to rocks; at depths of 20-600 ft. (6.1-182.8 m). *Distribution:* Central Gulf to Panama. *Remarks:* This locally common hermit crab is usually found in deeper water and is often trawled by shrimpers.

223. ***Hippa pacifica***

Pacific mole crab

Description: Two rounded lobes are located on the anterior margin of the carapace. The body color is variable, usually marbled with white markings against a gray, brown, or orange background. *Size:* Carapace length 0.5-0.75 in. (12-19 mm). *Habitat:* Burrows in sand; intertidal and subtidal to 75 ft. (22.9 m). *Distribution:* Central Gulf to Ecuador; and throughout the Indo-Pacific. *Remarks:* Unlike the closely related sand crabs (e.g. *Emerita*) that are suspension feeders, mole crabs of the genus *Hippa* are primarily detritivores, feeding on animal matter washed ashore by the tides.

224. ***Grapsus grapsus***

Sally lightfoot

Description: The adult body is a variegated red, brown and green; juveniles are usually dull gray to dark brown. *Size:* Carapace length 2-3.5 in. (50-87 mm). *Habitat:* On rocks and cliffs by the water's edge, particularly around offshore islands. *Distribution:* Central Gulf to Chile. *Remarks:* Large aggregations occur in some localities. These fast-moving, agile crabs are able to jump from rock to rock. They often share the same territory with sea lions.

225. ***Percnon gibbesi***

Spray crab

Description: The carapace and limbs are flattened. There are numerous spines on the legs. *Size:* Carapace length 1.5-2 in. (38-50 mm). *Habitat:* In narrow, rocky cliff crevices near the high water line. *Distribution:* Lower Gulf to Chile. *Remarks:* This is an agile and fast-moving crab.

226. ***Uca princeps***

Princely fiddler crab

Description: The large claw of the male is pustulated. Body is tan to bluish-brown. *Size:* Carapace length 1.25-2 in. (32-50 mm). *Habitat:* Mud flats and estuaries. *Distribution:* Throughout the Gulf to Peru. *Remarks:* Aggregations of hundreds of these crabs, even thousands, are not uncommon. This is the largest fiddler crab in the Gulf.

227. ***Callinectes arcuatus***

Pacific blue crab, blue swimming crab

Description: Can be separated from similar *Callinectes bellicosus* by the presence of fine granules on the carapace (smooth in *C. bellicosus*). Also, the walking legs of *C. arcuatus* are bluer than *C. bellicosus*. *Size:* Carapace width (including lateral spines) 4-5 in. (100-125 mm). *Habitat:* On sand and mud substrates in estuaries, mangrove swamps, and tidal flats; intertidal to 150 ft. (46 m). *Distribution:* Throughout the Gulf to Peru; outer coast of Baja California from Cabo San Lucas to southern California. *Remarks:* This is one of the most common swimming crabs in the Gulf. It often buries in sand or mud. Older individuals that have not molted for many months usually have barnacles attached to their carapaces.

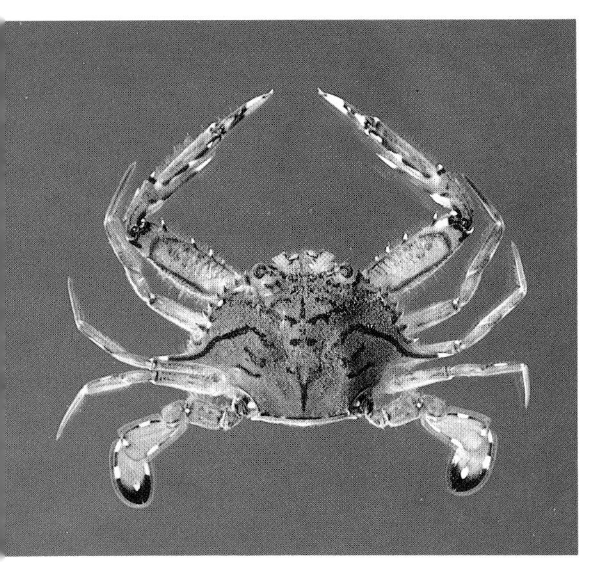

228. ***Portunus iridescens***

Iridescent swimming crab

Description: The carapace is light brown with reddish striations. The swimming "paddles" of the last pair of legs are tipped with red and bordered with small white spots. *Size:* Carapace length 3-4 in. (75-100 mm). *Habitat:* Offshore on sandy bottoms; at depths of 50-300 ft. (15.2-91.4 m). *Distribution:* Central Gulf to Colombia. *Remarks:* Large numbers of these swimming crabs are often captured in shrimp trawls.

229. ***Aethra scruposa scutata***

Walking rock crab

Description: The carapace is flattened and cryptic, resembling rocks. *Size:* Carapace length 3-5 in. (75-125 mm). *Habitat:* Among, and under rocks, and rubble; intertidal and subtidal to 60 ft. (18.3 m). *Distribution:* Central Gulf to Ecuador. *Remarks:* These uncommon crabs are difficult to see underwater. They so resemble flat rocks that they may be noticed only when they move.

230. ***Eriphia squamata***

Redeyed pebble crab

Description: The carapace is covered with small tubercles, which are pronounced on the claws. *Size:* Carapace length 1-2 in. (25-50 mm). *Habitat:* In crevices and under rocks; intertidal and subtidal to 50 ft. (15.2 m). *Distribution:* Throughout the Gulf to Ecuador. *Remarks:* This is one of the most common crabs in the Gulf.

231. ***Panopeus purpureus***

Pacific pebble crab

Description: The carapace color varies from white to reddish brown or gray. The claws are dark brown or gray, with white tips. *Size:* Carapace length 0.75-2 in. (19-50 mm). *Habitat:* Under and among rocks; intertidal and subtidal to 30 ft. (9.1 m). *Distribution:* Throughout the Gulf to Peru. *Remarks:* This crab is locally common, particularly in the mid and low intertidal zones.

232. ***Platypodia rotundata***

Orange pebble crab

Description: The carapace is red to orange. The eye-stalks are white and the tips of the claws are dark gray to black. The margin of the carapace is sharp-edged. *Size:* Carapace length 0.5-0.75 in. (12-19 mm). *Habitat:* Under subtidal rocks; at depths of 10-50 ft. (3-15.2 m). *Distribution:* Throughout the Gulf to Ecuador. *Remarks:* This uncommon crab is occasionally found living commensally among colonial anemones.

233. ***Carpilodes cinctimanus***

Redbanded crab

Description: Adults are brown with red bands on the legs and claws; juveniles are white to pink. *Size:* Carapace length 1-1.5 in. (25-38 mm). *Habitat:* Under subtidal rocks; at depths of 5-30 ft. (1.5-9.1 m). *Distribution:* Throughout the Gulf to Ecuador; also tropical west Pacific. *Remarks:* Although this uncommon species is usually found under rocks, it occasionally lives among the branches of the stony coral, *Pocillopora,* and under dead coral slabs.

234. ***Cycloxanthops vittatus***

Chameleon crab

Description: The carapace color is highly variable, with at least three typical forms: (1) red spots against a white background; (2) black spots on light brown background; and (3) dark brown with a white edge on the anterior part of the carapace. *Size:* Carapace length 0.5-1 in. (12-25 mm). *Habitat:* Under intertidal and subtidal rocks; to 30 ft. (9.1 m). *Distribution:* Central Gulf to Ecuador. *Remarks:* The red-spotted color form of this uncommon crab appears to be the one most often seen.

235. ***Quadrella nitida***

Sea fan crab

Description: The carapace and legs are white, with enlarged red to reddish brown chelipeds. *Size:* Carapace length 0.25-0.37 in. (6-9 mm). *Habitat:* Commensal on gorgonians; at depths of 40-150 ft. (12.2-46 m). *Distribution:* Central Gulf to Ecuador. *Remarks:* This common crab is mostly found living among the branches of *Muricea, Eugorgia,* and *Pacifigorgia.* As many as 10 individuals may exist on a single gorgonian. Each individual clings tenaciously to gorgonian branches and is difficult to remove without tearing legs and claws.

236. ***Stenorhynchus debilis***

Panamic arrow crab

Description: The carapace is pink to reddish brown, streaked with yellowish lines. The legs are long and spider-like. *Size:* Carapace length 0.25-0.75 in. (6-19 mm). *Habitat:* In crevices, at the base of cliffs; at depths of 5-200 ft. (1.5-61 m). *Distribution:* Throughout the Gulf to Chile. *Remarks:* In the Gulf this species appears to be seasonal, reaching its greatest size from May to July. It is occasionally found under the tentacles of the tube anemone, *Pachycerianthus*.

237. ***Epialtoides paradigmus***

Cryptic spider crab

Description: The carapace color is variable, from rusty-brown to reddish-yellow, with irregular markings. The arms are relatively long. It possesses a long flat rostrum and five free abdominal segments. *Size:* Carapace length 0.25-0.50 in. (6-12 mm). *Habitat:* Under rocks and in sponges; intertidal zone to 40 ft. (12.2 m). *Distribution:* Throughout the Gulf to Mazatlán. *Remarks:* This small crab often blends in remarkably well with its surroundings.

238. ***Herbstia camptacantha***

Flat spider crab

Description: The carapace and legs are covered with short spines. *Size:* Carapace length 0.5-1.5 in. (12-38 mm). *Habitat:* Among subtidal rocks and rubble; at depths of 3-200 ft. (.9-61 m). *Distribution:* Throughout the Gulf to Panama. *Remarks:* This crab is uncommon inshore, but relatively common offshore. It is often trawled by shrimpers.

239. ***Ala cornuta***

Gulf decorator crab

Description: The carapace color is highly variable, from brown to dull yellow or reddish, with occasional dark mottling. *Size:* Carapace length 0.75-2.5 in. (19-63 mm). *Habitat:* Under intertidal and subtidal rocks and in crevices; at depths of 5-60 ft. (1.5-18.3 m). *Distribution:* Throughout the Gulf to Peru. *Remarks:* This crab usually covers its carapace with various organisms such as algae, anemones, and sponges, which adhere to its hook-like dorsal spines.

240. ***Stenocionops angusta***

Inflated spider crab

Description: Similar to *Ala cornuta* but larger. The carapace is textured with short spines and the long chelae have black tips. *Size:* Carapace length 1-2.5 in. (25-63 mm). *Habitat:* On subtidal rocks and sandy bottoms; at depths of 10-200 ft. (3-61 m). *Distribution:* Throughout the Gulf, and west Baja California from Cabo San Lucas north to at least Scammon's Lagoon. *Remarks:* This uncommon species is occasionally taken by shrimp trawlers or by scuba divers on rocks near sand.

241. ***Hepatus lineatus***

Panamic calico crab

Description: The tumid carapace is covered with orange-brown spots of various sizes. *Size:* Carapace length 2-5 in. (50-125 mm). *Habitat:* Normally burrows under the sand; low intertidal to 250 ft. (76.2 m). *Distribution:* Throughout the Gulf to Panama. *Remarks:* Large numbers of these crabs are often taken by shrimp trawlers.

242. ***Hepatus kossmani***

Striped box crab

Description: Similar to *Hepatus lineatus,* but striped rather than spotted. *Size:* Carapace length 2-3.25 in. (50-82 mm). *Habitat:* Burrows under the sand; low intertidal to 400 ft. (122 m); most common offshore. *Distribution:* Throughout the Gulf to Ecuador. *Remarks:* Like *Hepatus lineatus,* this species is occasionally trawled by shrimpers but is not nearly as common.

243. ***Calappa convexa***

Shame faced crab

Description: The carapace and claws are highly tuberculate. *Size:* Carapace length 2-4 in. (50-100 mm). *Habitat:* On sand, offshore; at depths of 60-400 ft. (18.3-122 m). *Distribution:* Throughout the Gulf to Ecuador. *Remarks:* Large numbers of this species are often captured by shrimp trawlers.

244. ***Dromidia larraburei***

Sponge crab

Description: The carapace is tan and covered with a thick coat of very fine brown hair. *Size:* Carapace length 1-2 in. (25-50 mm). *Habitat:* Among rocks; low intertidal and subtidal to at least 60 ft. (18.3 m). *Distribution:* Throughout the Gulf to Peru. *Remarks:* This crab usually carries a sponge on its back, but occasionally will be found with a sea anemone or tunicate.

245. ***Hypoconcha lowei***

Shell crab

Description: Carapace and legs are covered with fine brown hairs. *Size:* Carapace length 1-2.5 in. (25-63 mm). *Habitat:* Among rubble and on patch reefs; low intertidal to 300 ft. (91.4 m). *Distribution:* Throughout the Gulf to Ecuador. *Remarks:* This relatively common crab normally carries an empty clam valve on its back. It is often captured by shrimp trawlers.

PHYLUM ECHINODERMATA
Class Asteroidea (Seastars)

246. ***Astropecten armatus***

Spiny sand star

Description: The flattened, star-shaped body has long arms, wide at the base, and tapering to the ends. The body is covered with paxillae. The tube feet do not have suckers. *Size:* Diameter 5-14 in. (126-360 mm). *Habitat:* Burrows just under the surface in sand and soft mud; low-intertidal zone to at least 525 ft. (160 m). *Distribution:* Throughout the Gulf to Peru, outer coast of Baja California from Cabo San Lucas north to southern California. *Remarks:* This species may be the same as *Astropecten verrilli*. It feeds on a variety of benthic invertebrates and also scavenges. Its color varies from gray to pink or lavender to orange and scarlet. It is often captured by shrimp trawlers.

247. ***Tethyaster canaliculatus***

Channeled sea star

Description: This sea star has a large disc and broad, flattened, tapering arms. Both the upper and lower marginal arm plates are large and conspicuous. The arms are generally unequal in length. The tube feet do not have suckers. *Size:* Diameter 6-20 in. (150-500 mm). *Habitat:* Burrows in subtidal sand and mud; at depths of 20-585 ft. (6-178 m). *Distribution:* Central Gulf to Panama, outer coast of Baja California north to Bahía Vizcaino. *Remarks:* This sea star is frequently encountered in shrimp trawls in the midriff area of the Gulf.

248. ***Amphiaster insignis***

Spiny sea star

Description: The body is ornately-spined with a flat disc and broad, triangular arms. Large, smooth conical spines cover both upper and lower surfaces and large granular plates border the upper and lower arm margins. *Size:* Diameter 3-7 in. (76-174 mm). *Habitat:* On sand and mud; intertidal zone to 420 ft. (128 m). *Distribution:* Central Gulf to Panama. *Remarks:* A similar-appearing sea star, *Paulia horrida,* from the same family, is found outside the Gulf.

249. ***Nidorellia armata***

Chocolate chip sea star

Description: The body is pentagonal and somewhat inflated. Large oblong pedicellariae are scattered on both surfaces. Ambulacral grooves containing suckered tube feet are often visible on the upturned arm tips when viewed from above. *Size:* Diameter 3-6 in. (76-150 mm). *Habitat:* Rocky substrates; low-intertidal zone to 240 ft. (73 m). *Distribution:* Throughout the Gulf to northern Peru and the Galápagos. *Remarks:* A common rocky shore inhabitant that feeds on both benthic algae and sessile invertebrates. The commensal shrimp, *Periclimenes soror,* is associated with this species in deeper water.

250. ***Pentaceraster cumingi* (formerly *Oreaster occidentalis*)**

Panamic cushion star

Description: A beautiful cushion star with a hard, inflated body wall. Stout, immobile spines stud the upper surface. There are no spines on the lower surface or on the marginal arm plates. The suckered tube feet are usually well-hidden within the deep ambulacral grooves. *Size:* Diameter 4-7 in. (100-174 mm). *Habitat:* Rocky substrates and patch reefs; low-intertidal zone to 600 ft. (183 m). *Distribution:* Throughout the Gulf to northern Peru and the Galápagos; also in Hawaii. *Remarks:* The commensal shrimp, *Periclimenes soror,* is associated with this sea star. Large aggregations are reported in mid-summer months in the central Gulf, particularly in Baja California.

251. ***Asteropsis carinifera***

Keeled sea star

Description: A flattened sea star, especially on the lower (oral) surface, with a flexible body and arms that are triangular in cross-section. A smooth skin covers the body plates, giving the animal a wet or slimy appearance. A ridge of spines runs down the center of each arm. *Size:* Diameter 4-10 in. (100-250 mm). *Habitat:* Rocky reefs and the undersides of rock slabs; intertidal zone to 120 ft. (36.5 m). *Distribution:* Central Gulf to Panama and the Galápagos; throughout the Indo-Pacific. *Remarks:* It is one of the most common and widely distributed sea stars of the Indo-Pacific, where the predatory shrimp *Hymenocera* reportedly feeds on it.

252. ***Leiaster teres***

Smooth sea star

Description: This sea star is distinguished by its thick, shiny skin, a small disc, and long rounded arms. It has three rows of overlapping plates on the upper arm surface and two rows on the sides of the arm. There are numerous groups of small finger-like respiratory papillae on the upper (aboral) surface. *Size:* Diameter 3.5-17 in. (88-425 mm). *Habitat:* In rocky and muddy substrates; low-intertidal zone to 187 ft. (57 m). *Distribution:* Central Gulf to Panama. *Remarks: Leiaster* has a characteristic glossy or shiny appearance that distinguishes it from most other Gulf sea stars.

253. ***Pharia pyramidata***

Pyramid sea star

Description: This sea star has a small disc relative to its long, bluntly tapering arms. It is hard and smooth-bodied and has very narrow ambulacral grooves guarded by flattened spines. Rows of plates on the tops and sides of the arms give it an angular appearance. The tube feet have weak suckers. *Size:* Diameter 5.5-12 in. (140-300 mm). *Habitat:* Rocky substrates; low-intertidal zone to 456 ft. (139 m). *Distribution:* Throughout the Gulf to northern Peru and the Galápagos; also on some of the west Baja California islands. *Remarks:* This common species resembles and almost always occurs with *Phataria unifascialis,* from which it differs most noticeably in size and coloration. It feeds on algae and soft-bodied sessile invertebrates.

254. ***Phataria unifascialis***

Tan sea star

Description: The overall color varies from brown to reddish brown. The body is slender and stiff, with a very small disc. The long, rounded, tapering arms have a smooth granulated surface. *Size:* Diameter 4-8 in. (100-200 mm). *Habitat:* Rocky substrates; low-intertidal zone to 450 ft. (139 m). *Distribution:* Throughout the Gulf to northern Peru and the Galápagos. *Remarks:* It is the most abundant shallow-water sea star in rocky habitats in the Gulf. Like *Pharia,* it is frequently associated with coralline algae and seems to be primarily herbivorous.

255. ***Tamaria stria***

Tamarisk sea star

Description: The arms are long and cylindrical. Each arm has rows of pores, where the fingerlike gill papillae protrude, with 6-10 holes per pore area. The body is covered with granules and the disc is small. *Size:* Diameter 3-7 in. (76-174 mm). *Habitat:* Rock and rubble substrates; at depths of 40-165 ft. (12-50 m). *Distribution:* Islands from the central Gulf to Colombia. *Remarks:* There are two known species of *Tamaria* in the eastern Pacific Ocean, but they apparently are rare. Although this sea star has been found around offshore islands of the central Gulf, it has been more frequently encountered in the southern Gulf at somewhat greater depths.

256. ***Mithrodia bradleyi***

Bradley's sea star

Description: This sea star has the upper surface covered with short rounded spines, and has wide ambulacral grooves bearing suckered tube feet. *Size:* Diameter 3.5-14 in. (140-350 mm). *Habitat:* Rocky substrates; low-intertidal zone to 165 ft. (50 m). *Distribution:* Central Gulf to the Galápagos, plus one record from Punta Eugenia, outer Baja California. *Remarks:* This conspicuous species may prove to be identical to *Mithrodia clavigera* of the Indo-Pacific. It has a tendency to disintegrate rapidly under stress, shedding arms at the disc when it is removed from the water.

257. ***Acanthaster ellisii***

Panamic crown-of-thorns

Description: This spiny, multi-armed sea star has 13-16 arms that are short in relation to the large disc. The body is quite flexible, with wide ambulacral grooves. The numerous tube feet have suckers. *Size:* Diameter 4-16 in. (100-400 mm). *Habitat:* Rocky reefs, particularly on offshore islands where rich coral and gorgonian growth occurs; low-intertidal zone to 150 ft. (46 m). *Distribution:* Central Gulf to Clarion Island; may occur in the Galápagos and northern Peru. *Remarks:* This sea star may be the same as the notorious Indo-West Pacific *Acanthaster planci,* which periodically decimates coral reefs. A commensal shrimp, *Periclimenes soror,* is found among its numerous sharp and venomous spines.

258. ***Echinaster tenuispina* (formerly *Othilia tenuispina*)**

Thin spined sea star

Description: The overall color varies from yellow-ochre to bluish-gray, or dull reddish-brown. There are purple spots on the tips of the arms. *Size:* Diameter 1.5-8 in. (38-200 mm). *Habitat:* Rocky substrates with algae and sand; mid-intertidal zone to 240 ft. (73 m). *Distribution:* Throughout the Gulf and around the tip of outer Baja California, north to Scammon's Lagoon. *Remarks:* This once-abundant Gulf species suffered a drastic population decline in 1978 from a widespread infection of unknown origin, and it still has not completely recovered.

259. ***Heliaster kubiniji***

Gulf sun star

Description: Usually has 20-25 stout arms that are rounded at the tips. Some individuals, particularly juveniles, have a banded appearance. *Size:* Diameter 2.5-8 in. (63-200 mm). *Habitat:* Rocks, boulders, and reefs; mid-intertidal zone to 120 ft. (37 m). *Distribution:* Throughout the Gulf to Nicaragua, and around the tip of Baja California north to Bahía Magdalena; scattered occasional records exist from northwest Baja California and southern California. *Remarks:* This species was once the most abundant intertidal sea star in the Gulf but in 1978 populations were almost completely wiped out by a widespread infection, although it is now increasing in numbers. It feeds primarily on barnacles and mussels but is also a scavenger.

260. ***Astrometis sertulifera***

Fragile rainbow star

Description: A striking, soft-bodied and somewhat "slimy" sea star, with a small disc and long tapering arms. The well-separated, tapering spines are surrounded at their bases by rosettes of pedicellariae. *Size:* Diameter 3-8 in. (76-200 mm). *Habitat:* Rocks and patch reefs; low-intertidal zone to 512 ft. (156 m). *Distribution:* Throughout the Gulf to northern Peru and the Galápagos; around the tip of Baja California north to Vancouver Island, Canada. *Remarks:* This species has become rare in the Gulf since 1978, when its populations were drastically reduced by a widespread infection. *Astrometis sertulifera* can trap zooplankton with its pedicellariae.

261. ***Rathbunaster californicus***

Rathbun's sea star

Description: The number of arms varies from 12 to 20 (usually 17). The upper body plates have needle-like spines encircled by a conspicuous wreath of short, crossed pedicellariae sitting on a heavy sheath. *Size:* Diameter 5-12 in. (126-300 mm). *Habitat:* Sand and mud, occasionally at depths of 100 ft.-2500 ft. (30.5-768 m). *Distribution:* Tip of Baja California to northern California (distribution south of the Gulf is unrecorded). *Remarks:* The rays of this species are quite flexible due to the absence of a connected skeleton on the upper surface. The conspicuous wreaths of pedicellariae are used to capture small prey.

Class Ophiuroidea (Brittle Stars)

262. ***Astrocaneum spinosum***

Spiny basket star

Description: Each arm has five to six major branches at its base, with the branching pattern changing from dichotomous (Y-shaped) near the disc to alternate (left-right) towards the ends. The arms have a series of five to six spines down their midline, otherwise they are granular. *Size:* Diameter 5-10 in. (125-250 mm). *Habitat:* On gorgonians, over rubble, sand and mud; at depths of 30-600 ft. (9-183 m). *Distribution:* Throughout the Gulf to Panama and around the tip of Baja California to Punta Eugenia; Japan. *Remarks:* This basket star is most often associated with gorgonians.

263. ***Ophiocoma aethiops***

Black spiny brittle star

Description: The arm side spines are stout, as long as the diameter of the stout arms, and perpendicular to the arm axis. The disc is covered with fine granules and often has a scalloped appearance around the edge due to "pouching" between the arm insertions. *Size:* Diameter 3.5-19.5 in. (88-494 mm). *Habitat:* Under large boulders, rocky reefs and crevices; mid-intertidal zone to 100 ft. (30.5 m). *Distribution:* Throughout the Gulf to northern Peru and the Galápagos; around the tip of Baja California north to Bahía Vizcaino. *Remarks:* This common brittle star is very active and thought to be a carrion and opportunistic feeder.

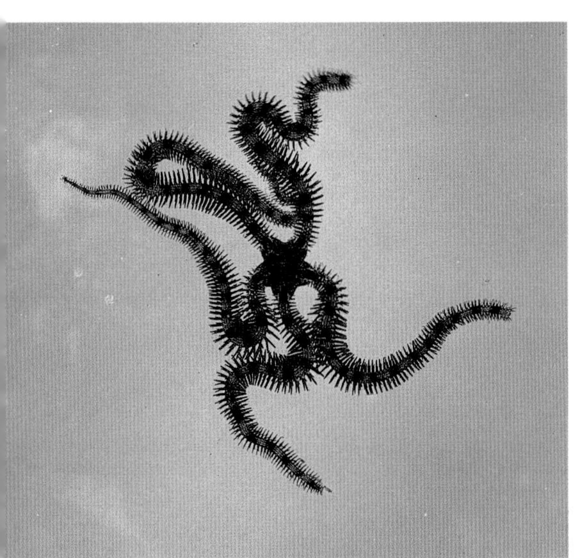

264. ***Ophiocoma alexandri***

Alexander's spiny brittle star

Description: The long arms are relatively slender and flattened, with five to seven long spines on each lateral plate. The arms are banded, but only on the upper surface. The disc has a sandpaper-like feel. *Size:* Diameter 3.5-17.5 in. (88-438 mm). *Habitat:* On rocky reefs, and under boulders on sand; intertidal zone to 230 ft. (70 m). *Distribution:* Throughout the Gulf to Colombia and the Galápagos. *Remarks:* This abundant species can attain sizes comparable to *Ophiocoma aethiops* but is normally smaller. It is more common than *O. aethiops* in the Gulf. These two species overlap in distribution as well as in habitat.

265. ***Ophionereis annulata***

Ringed brittle star

Description: The disc is covered with small overlapping scales which increase in size towards the periphery and partly cover the first upper arm plate. The arms are long and slender, with three spines on each lateral arm plate. The arms are banded on both upper and lower surfaces. *Size:* Diameter 3-12.5 in. (50-315 mm). *Habitat:* Under rocks and in crevices, also on sponges, corals, and sand; intertidal zone to 90 ft. (27.4 m). *Distribution:* Throughout the Gulf to Ecuador and the Galápagos, and around the tip of Baja California north to southern California (Pt. Conception). *Remarks:* This is the most common intertidal brittle star in the Gulf. It moves by "stepping" with its tube feet rather than by using entire arm movements.

266. ***Ophioderma panamense***

Panamic brittle star

Description: A smooth, slate-colored brittle star with slender, slightly flattened arms and very short, equal-sized arm spines. The upper arm plates are usually unbroken. The arms are always banded, especially at the tips. The disc has thickened, granular skin. *Size:* Diameter 2.5-10 in. (63-250 mm). *Habitat:* Under rocks and in crevices; intertidal zone to 65 ft. (20 m). *Distribution:* Throughout the Gulf to northern Peru and the Galápagos; around the tip of Baja California north to southern California. *Remarks:* This brittle star is particularly abundant in the intertidal zone. There are three color phases of this species, each associated with a particular habitat.

267. ***Ophioderma variegatum***

Multicolored brittle star

Description: The upper arm plates of this brittle star are undivided, rectangular, and often have a wavy leading edge. The arms are conspicuously banded. There are spots on the disc, sometimes irregular in shape and number and sometimes numbering five (one between each arm). The color varies, but the disc is usually red. *Size:* Diameter 3.5-7 in. (86-175 mm). *Habitat:* Under rocks on rubble, mud, and sand; intertidal zone to 360 ft. (110 m). *Distribution:* Throughout the Gulf to Panama and the Galápagos, and around the tip of Baja California to San Diego, California.

Class Echinoidea (Sea Urchins)

268. ***Eucidaris thouarsii***

Slate pencil urchin

Description: The uppermost plates form a regular, five-rayed star pattern. It has 10 vertical rows of 5-8 coarsely-sculptured, heavy spines each surrounded by 2 rings of short, flat spines. *Size:* Test diameter 1.25-2.75 in. (32-70 mm), longest spines to 2 in. (50 mm). *Habitat:* In reef depressions, under rocks, and in crevices; mid-intertidal zone to 490 ft. (150 m). *Distribution:* Throughout the Gulf to Ecuador and the Galápagos and around the tip of Baja California to southern California. *Remarks:* This species is common throughout the Gulf but rare north of central west Baja California. Pencil urchins are almost always found with encrusting organisms.

R.Brusca

269. ***Centrostephanus coronatus***

Crowned sea urchin

Description: The spines are brittle and annulated and, in juveniles, are always banded. Flat spinelets surround the uppermost plates on the test and five clusters of small spines surround the mouth. *Size:* Test diameter 1.25-2.5 in. (32-63 mm), longest spines over 5 in. (125 mm). *Habitat:* Rock, reef, and sand, especially sheltered substrates; low-intertidal zone to 410 ft. (125 m). *Distribution:* Throughout the Gulf to northern Peru and the Galápagos, and around the tip of Baja California north to southern California. *Remarks:* Although most sea urchins are herbivores, this nocturnal species is mostly carnivorous. It has a summer monthly reproductive rhythm that is synchronized with lunar cycles.

270. ***Lytechinus pictus* (formerly *L. anamesus*)**

Embroidered sea urchin

Description: The test is thin, flattened underneath and distinctly domed on top. The spines are short (about one-quarter the diameter of the test), thick, and blunt, sometimes banded. The color of the test, presence of banded spines, and length of spines are all quite variable. *Size:* Test diameter 0.37-1.75 in. (9-44 mm), longest spines 1 in. (25 mm). *Habitat:* Sand, mud, rocks, and eelgrass; low-intertidal zone to 985 ft. (300 m). *Distribution:* Throughout the Gulf to Ecuador and around the tip of Baja California, north to Monterey, California. *Remarks:* This urchin is nocturnal, often buries in sand during the day, and sometimes covers itself with bits of algae and other debris by using its long tube feet.

271. ***Toxopneustes roseus***

Flower sea urchin

Description: The spines are very short. Huge globiferous pedicellariae, which are comonly held open, give this species the appearance of being covered with flowers or bows. *Size:* Test diameter 1.75-5 in. (44-125 mm), longest spines to 0.5 in. (12 mm). *Habitat:* Rocks, coral, sand, and mud; low-intertidal zone to 80 ft. (24 m). *Distribution:* Throughout the Gulf to Ecuador and the Galápagos, and around the tip of Baja California to Bahía Tortugas. *Remarks:* The large rosy-colored pedicellariae, which are venomous. It sometimes covers itself with bits of algae and shell debris.

272. *Echinometra vanbrunti*

Purple sea urchin

Description: Overall spine color is dark purple. The spines are sturdy, of medium length, and taper to a sharp point. *Size:* Test diameter 1.5-2.75 in. (38-70 mm), longest spines 0.37-2.5 in. (9-63 mm). *Habitat:* Rock reefs and cliffs in crevices and cavities; mid-intertidal zone to 175 ft. (53.4 m). *Distribution:* Throughout the Gulf to Peru and the Galápagos; tip of Baja California north to Bahía Magdalena. *Remarks:* This is the most common sea urchin in the Gulf and throughout the Panamic region. It is a rock-borer, often occupying its own cavity in a rock. It is preyed upon by various sea stars, and the finescale triggerfish, *Balistes polylepis*.

273. *Asthenosoma varium*

Galloping urchin

Description: All spines are hollow and many are highly modified. Some spines look like clubs and are covered with a jelly-filled skin that gives the urchin a neutral buoyancy. Five double rows of dorsal spines in the interambulacra have blue venom sacs. *Size:* Test diameter 1.5-8 in. (38-200 mm). *Habitat:* Often on mud, but also on rocky bottoms; intertidal zone to 930 ft. (285 m). *Distribution:* Known throughout the Indo-West Pacific, it has recently been found on the El Bajo seamount off La Paz. *Remarks:* The glandular "flotation" sacs on the upper surface, in combination with the hooved spines on the lower surface, gives this urchin a high degree of mobility, allowing it to move very swiftly. The spines are extremely toxic and can inflict deadly stings.

274. *Encope grandis*

Large holed sand dollar

Description: The large posterior lunule (hole) varies from narrowly elongated to circular. The posterior margin is generally flattened, and the overall test shape is often quite irregular. Color varies from dull purple to black and is often darker around the edges than in the center. *Size:* Test length 1.5-5 in. (38-125 mm). *Habitat:* On sand flats; low-intertidal zone to 150 ft. (46 m). *Distribution:* Throughout the Gulf and around the tip of Baja California north to Bahía Magdalena. *Remarks:* This sand dollar is often found in large beds with another Gulf keyhole urchin, *Encope micropora*.

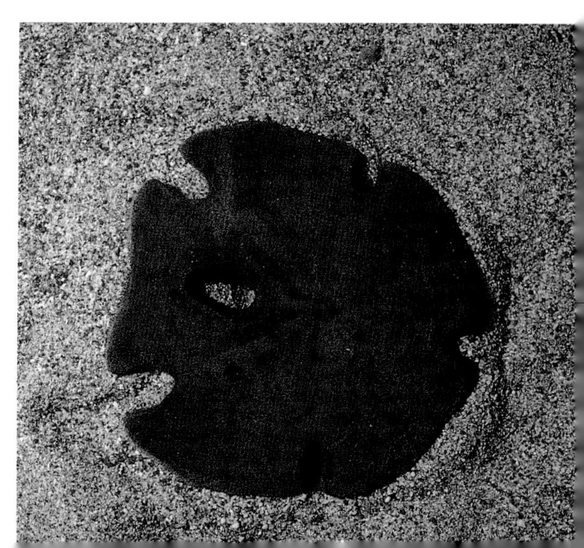

275. ***Clypeaster europacificus***

Giant sand dollar

Description: This sand dollar has a definite pentagonal shape, with rounded angles. The interradial margins are usually concave and the petals stand out clearly. The test is not very stout and injuries to the margin often result in deformities. The color of this species varies considerably, especially with growth. Color generally deepens with age and can be tan, brown, red-violet, violet, or brilliant green. *Size:* Test diameter 3-8 in. (75-200 mm). *Habitat:* On sand; shallow subtidal to 330 ft. (100.5 m). *Distribution:* Throughout the Gulf to Ecuador and the Galápagos. *Remarks:* The young of this species are quite circular and may be confused with *Clypeaster rotundus.* Older individuals can be easily distinguished by the pentagonal shape.

276. ***Mellita longifissa***

Long slit keyhole sand dollar

Description: Test is almost circular, with narrowly-slit lunules that are not always closed along the margin. The posterior keyhole lunule is also a long narrow slit. *Size:* Test length 0.75-3 in. (19-75 mm). *Habitat:* Sand flats and patches of sand; mid-intertidal to 195 ft. (60 m). *Distribution:* Throughout the Gulf to Ecuador, and around the tip of Baja California north to Cedros Island. *Remarks:* This is an abundant species, particularly in the upper Gulf.

277. ***Agassizia scrobiculata***

Grooved heart urchin

Description: The color varies from pale brown to white. The test is oval and inflated, with short spines. Both ambulacral areas and bands of spines are quite obvious. *Size:* Test length 0.37-2.25 in. (9-56 mm). *Habitat:* Burrows in sand, mud, and sometimes under rocks; low-intertidal zone to 250 ft. (76.2 m). *Distribution:* Throughout the Gulf to northern Peru and the Galápagos, and around the tip of Baja California north to Bahía Magdalena. *Remarks:* Common in the Gulf, but less so in other parts of its range. It is often captured in large numbers by shrimp trawlers.

278. ***Lovenia cordiformis***

Heart urchin, sea porcupine

Description: The color varies in this species from off-white to pale brown, to purple. The innermost band of spines on the central upper surface is conspicuous. *Size:* Test length 0.37-3 in. (9-75 mm); posterior dorsal spines 4 in. (100 mm). *Habitat:* Burrows in sand, mud, and silt; low-intertidal and subtidal to 656 ft. (200 m). *Distribution:* Throughout the Gulf to northern Peru and the Galápagos, and around the tip of Baja California north to Santa Barbara, California; also in Hawaii. *Remarks:* This fairly common heart urchin feeds both on the detritus within its burrow and detrital matter pulled from the surface with its long tube feet. The long, posterior dorsal spines of this species can be held erect and used for defense.

Class Holothuroidea (Sea Cucumbers)

279. ***Psolidium dorsipes***

Plated sea cucumber

Description: The upper surface is covered with large overlapping scales. A few modified tube feet are found on these scales. The scales around the mouth and anus are very large and the mouth is directed upwards. The lower body surface is flattened into a sole with three bands of tube feet. *Size:* Body length 0.25-1 in. (6-25 mm). *Habitat:* Patch reefs, or rocks and algae; at depths of 36-360 ft. (11-110 m). *Distribution:* Central Gulf to Panama. *Remarks:* Very little seems to be known about this small, uncommon species. It is occasionally captured by shrimp trawlers.

280. ***Holothuria impatiens* (formerly *Brandtothuria*)**

Brown spotted sea cucumber

Description: The body is slender and often distinctly bottle-shaped with a long "neck." It is knobby and sandy to the touch. The tube feet, which form distinct rows, are large on the underside of the body and small on top. *Size:* Body length 1.75-6.5 in. (44-162 mm). *Habitat:* Rocky reefs and boulders, coral, and sand; intertidal zone to 150 ft. (46 m). *Distribution:* Throughout the Gulf to Ecuador and the Galápagos, and around the tip of Baja California north to Bahía Rosario; circumtropical. *Remarks:* The abundance of this species is not related to food but to microhabitat availability.

281. ***Holothuria lubrica* (formerly *Selenkothuria lubrica*)**

Sulfur sea cucumber

Description: This species varies in color from dull gray or brown to black. The body is soft-skinned, and has 20 large tentacles around the mouth. Normally, small papillae cover the upper body surface and numerous large, soft, sulfur-colored tube feet cover the underside. *Size:* Body length 2-6.5 in. (50-162 mm). *Habitat:* Rocks and crevices, also mud and sand; mid-intertidal zone to 180 ft. (55 m). *Distribution:* Throughout the Gulf to Ecuador and the Galápagos, and around the tip of Baja California north to Punta Eugenia. *Remarks:* This abundant Gulf species congregates in areas where there are currents, feeding more or less continuously day and night.

282. ***Isostichopus fuscus***

Giant sea cucumber

Description: A large, heavy sea cucumber with thickened flanks and a well-developed flattened bottom. The skin is smooth and leathery to the touch. The mouth is directed downwards and circled by 20 tentacles. There are up to 20 blunt, lateral warts and a similar number in rows on the upper body surface. *Size:* Body length 4-13 in. (100-325 mm). *Habitat:* Rocky reefs and boulders, occasionally mud and sand; low-intertidal zone to 200 ft. (61 m). *Distribution:* Throughout the Gulf to Ecuador and the Galápagos, and around the tip of Baja California north to Bahía Vizcaino. *Remarks:* This species grazes on the surface of rocks and reefs.

283. ***Euapta godeffroyi***

Sticky sea cucumber

Description: The length of this species is highly variable because of its ability to stretch its body. The mouth is terminal, with a crown of featherlike tentacles. This species lacks tube feet and the body musculature produces a "bubbly" appearance, particularly when contracted. *Size:* Body length 12-50 in. (30-120 cm). *Habitat:* Coral, rocky and patch reefs, under rocks; low-intertidal zone to 150 ft. (46 m). *Distribution:* Throughout the Gulf; an Indo-Pacific species. *Remarks:* This long, extensible sea cucumber clings to the skin when handled and fragments easily.

References

Barham, E., and J. Davies
1968. Gorgonians and water motion studies in the Gulf of California. Underwater Nat. 5(3):24-28.

Barham, E., R.W. Gowdy, and F.H. Wolfson
1973. *Acanthaster* (Echinodermata, Asteroidea) in the Gulf of California. Fish. Bull. U.S. Natl. Ocean Atmos. Admin. 71(4):927-942.

Bertsch, H.
1977-1978. The Chromodoridinae nudibranchs from the Pacific coast of America. Parts I-IV. Veliger 20:107-118 and 307-327; 21:70-86 and 236-250.
1979. Tropical faunal affinities of opisthobranchs from the Panamic province (eastern Pacific). The Nautilus 93(2-3):57-61.

Bertsch, H., and S. Johnson
1983. Zoogeografia comparativa de los opistobranquios (Mollusca: Gastropoda) con enfasis en la cuenca Pacifica (Hawaii y California): Composicion faunal, afinidades provinciales y densidad submareal. Ciencias Marinas 8(2):125-153.

Bertsch, H., and A. Kerstitch
1984. Distribution and radular morphology of various nudibranchs (Gastropoda: Opisthobranchia) from the Gulf of California, Mexico. Veliger 26(4):264-273.

Brusca, R.C., and B. Wallerstein
1979. Zoogeographic patterns of idoteid isopods in the northeast Pacific, with a review of shallow-water zoogeography for the region. Bull. Biol. Soc. Wash., 3:67-105.

Brusca, R.C.
1980. Common Intertidal Invertebrates of the Gulf of California. 2nd Ed., Univ. Ariz. Press, Tucson. 513 pp.

Brusca, R.C., and D.A. Thomson
1977. The Pulmo Reefs of Baja California — true coral reef formation in the Gulf of California. Ciencias Marinas 1(3):37-53.

Brusca, R.C., and M. Gilligan
1983. Tongue replacement in a marine fish (*Lutjanus guttatus*) by a parasitic isopod (Crustacea: Isopoda). Copeia, 3:813-816.

Caldwell, R., and H. Dingle
1976. Stomatopods. Sci. Amer. 234: pp. 80-89.

Case, T.J., and M.L. Cody
1983. Island Biogeography in the Sea of Cortez. Univ. of Calif. Press, Berkeley, Calif. 307-341.

Clark, H.L.
1948. A report on the Echini of the warmer eastern Pacific, based on the collections of the Velero III. Allan Hancock Pacif. Exped. 8(5):225-351.

Corredor, D.
1978. Notes on the behavior and ecology of the new fish cleaner shrimp *Brachycarpus biunguiculatus* (Lucas) Crustaceana 35(1): 35-40.

Dana, T., and A. Wolfson
1970. Eastern Pacific crown-of-thorns starfish populations in the lower Gulf of California. Trans. S. Diego Soc. Nat. Hist. 16(4):83-90.

Deichmann, E.
1958. The Holothuroidea collected by the Velero III and IV during the years 1932-1954. Part 2.

Dickinson, M.G.
1945. Sponges of the Gulf of California. Allan Hancock Pac. Exped. 11(1):1-252.

Doderlein, L.
1916. Uber die Gattung *Oreaster* und verwandte. Zool. Jb. (syst.) 40:409-440.

Dungan, M.L., T.E. Miller, and D.A. Thomson.
1982. Catastrophic decline of a top carnivore in the Gulf of California rocky intertidal zone. Science. 216:989-991.

Durham, J.W., and J.L. Barnard
1952. Stony corals of the eastern Pacific collected by the Velero II and Velero IV. Allan Hancock Pacific Expeditions 16(1):1-110.

Durham, J.W., C.D. Wagner, and D.P. Abbott
1980. Echinoidea: The sea urchins. pp. 160-176, pls. 51-57. *in:* R.H. Morris, D.P. Abbott, and E.C. Haderlie (eds.), Intertidal Invertebrates of California. Stanford Univ. Press, Stanford, Calif.

Ebert, T.A., and D.M. Dexter
1976. Morphological comparison of *Mellita grantii* and *Mellita longifissa* (Echinodermata, Echinoidea, Family Scutellidae). Cienc. mar. 3(2):8-17.

Feder, H.M.
1980. Asteroidea: The sea stars. pp. 117-135, pls. 40-46. *in:* R.H. Morris, D.P. Abbott, and E.C. Haderlie (eds.), The Intertidal Invertebrates of California. Stanford Univ. Press, Stanford, Calif.

Farmer, W.M.
1963. Two new opisthobranch molluscs from Baja California. Trans. San Diego Soc. Nat. Hist. 13(6):81-84.

Fraser, C.M.
1948. Hydroids of the Allan Hancock Pacific Expeditions since March, 1938. Allan Hancock Expeditions. 4(5): 180-328.

Garth, J.S.
1960. Distribution and affinities of brachyuran Crustacea (Symposium: The biogeography of Baja

California and adjacent seas. Part II. Marine Biotas) Syst. Zool. 9:105-123.

Gotshall, D.W.
1982. Marine Animals of Baja California. Sea Challengers, Monterey, Calif. 112 pp.

Haig, J.
1978. Contribution toward a revision of the porcellanid genus *Porcellana*. Proc. Biol. Soc. Wash. 91(3):706714.

Harry, H.W.
1985. Synopsis of the supraspecific classification of living oysters (Bivalvia: Gryphaeidae and Ostreidae). Veliger. Vol. 28, No. 2.

Hendrickx, M., M.K. Wicksten, and A. van der Heiden
1983. Studies of the coastal marine fauna of southern Sinaloa, Mexico. IV. Preliminary report on Caridea. Proc. Biol. Soc. Wash. 96(1):pp. 67-78.

Hendrickx, M.
1984. The species of *Sicyonia* H. Milne-Edwards (Crustacea: Penaeoidea) of the Gulf of California, Mexico, with a key for their identification and a note on their zoogeography. Rev. Biol. Trop., 32(2):279-298.

Hochberg, G.G., and J.A. Couch
1971. Biology of cephalopods, pp. 220-230, in J.W. Miller and J.G. Vanderwalker, eds. Scientist-in-the-sea. Washington, D.C.: U.S. Dept. Interior.

Holthuis, L.B., and F. Alejandro Villalobos
1962. *Panulirus gracilis* Streets y *Panulirus inflatus* (Bouvier), dos especies de langosta de la costa del Pacifico de America. Anales del Inst. Biol. U.N.A.M. 32(1):250-276.

Holthuis, L.B., and H. Loesch
1967. The lobsters of the Galápagos Islands (Decapoda, Palinuridae) Crustaceana 12(2):215-222. *Scyllarides astori* descr. and illustr.

Hyman, L.H.
1955. The Invertebrates: Echinodermata, the coelomate *Bilateria*. Vol. IV. McGraw-Hill, New York. 763 pp.

Jangoux, M.
1980. Le genre *Leiaster* Peters (Echinodermata, Asteroidea: Ophidiasteridae). Rev. Zool. Afr. 94(1):87-108.

Keen, A.M.
1971. Sea Shells of Tropical West America (second edition). Stanford Univ. Press, Stanford, Calif. 1064 pp.

Kennedy, B., and J.S. Pearse
1975. Lunar synchronization of the monthly reproductive rhythm in the sea urchin *Centrostephanus coronatus* Verrill. J. Exp. Mar. Biol. Ecol. 17:323-331.

Kerstitch, A.
1980. Poisonous and Venomous Marine Animals. AV program. Educational Images.
1982. Cleaning Shrimps. Oceans No. 5, p. 41.
1982. Smashers and Spearers — A look at the remarkable mantis shrimps. Fresh and Marine Aquarium. Feb. pp. 40-42.

Maluf, L.Y.
1988. Composition and distribution of the Central Eastern Pacific echinoderms. Los Angeles Co. Mus., Technical Rept. No. 2, 242 pp.

Manning, R.
1967. Eastern Pacific expeditions of the New York Zoological Society. Stomatopod Crustacea. Zoologica 56:pp. 95-113.

Marcus, E., and E. Marcus
1970. Some gastropods from Madagascar and west Mexico. Malacologia 10(1):181-223.

McLaughlin, P.
1981. Revision of *Pylopagurus*. Bull. Marine Sci. vol. 31, No. 1 and 2. pp. 1-30, pp. 329-365.

Morris, R.H., D. Abbott, and E.C. Haderlie
1980. Intertidal Invertebrates of California. Stanford Univ. Press, Stanford, Calif. 690 pp.

Poorman, L.H.
1981. Comments on two misunderstood fusinids (Gastropoda: Fasciolariidae) from the tropical eastern Pacific. The Veliger Vol. 23(4):345-347.

Ristau, D.A.
1978. Six new species of shallow-water marine demosponges from California. Proc. Biol. Soc. Wash. 91(3):560-590.

Squires, D.F.
1959. Corals and Coral Reefs in the Gulf of California. Bull. Amer. Mus. Nat. Hist. 188(7):371-431.

Steinbeck, J., and E.F. Ricketts
1941. Sea of Cortez. A Leisurely Journey of Travel and Research. Viking Press, New York. 598 pp.

Stephenson, W.
1967. A comparison of Australian and American specimens of *Hemisquilla ensigera* (Owen, 1832) Proc. U.S. Nat. Mus. 120:1-18.

Thomson, D.A., and C. Lehner
1976. Resilience of a rocky intertidal fish community in a physically unstable environment. J. Exp. Mar. Biol. Ecol. 22:1-29.

Thomson, D.A., L.T. Findley, and A. Kerstitch
1987. Reef Fishes of the Sea of Cortez. University of

Arizona Press, Tucson, Arizona. 302 pp.
Vokes, E.H.
1984. Comparison of the Muricidae of the eastern Pacific and western Atlantic, with cognate species. Shells and Sea Life. 16(11): 210-216.
Wicksten, M.K.
1979. Zoogeographical affinities of the broken back shrimp (Caridea: Hippolytidae) of western South America. In Proc. Internat. Symp. on Mar. Biogeo. and Evol. in So. Hemisphere N.Z. DSIR Vol. 2:627-634.
1982. Two species of *Odontozona* (Decapoda: Stenopodidea) from the eastern Pacific. Journal of Crustacean Biology 2(1):130-135.
1983. A monograph on the shallow water caridean shrimps of the Gulf of California, Mexico. Allan Hancock Monographs in Marine Biology No. 13, pp. 1-59.
Wicksten, M.K., and M.C. Hendrickx
1985. New record of caridean shrimps in the Gulf of California, Mexico. Proc. Biol. Soc. Wash. 98(3):571-573.
Wiedenmayer, F.
1977. Shallow-water sponges of the western Bahamas. Basel and Stuttgart: Birkhauser Verlag. 287 pp.

Index to Scientific Names and Common Names of Major Groups